SCI

D0933607

There Are Giants
in the Earth

Also by Michael Grumley, with Robert Ferro

ATLANTIS: THE AUTOBIOGRAPHY OF A SEARCH

J599.8

MICHAEL GRUMLEY

There Are Giants in the Earth

Illustrations by the Author

Cop. a

Doubleday & Company, Inc.   Garden City, New York
1974

1408

0695

Copyright © 1974 by Michael Grumley
All Rights Reserved
Printed in the United States of America
First Edition

Library of Congress Cataloging in Publication Data

Grumley, Michael.
    There are giants in the earth.

    Bibliography: p. 147
    1. Sasquatch. 2. Yeti. 3. Primates—Legends and stories. I.
Title.
QL89.2.S2G78   599′.8
ISBN 0-385-07583-9
Library of Congress Catalog Card Number 73–81431

Excerpt from *Bigfoot* by John Napier. Copyright © 1972 by John
Napier. Reprinted by permission of E. P. Dutton & Co. Published
by Jonathan Cape, Ltd., London.

Portions of lyrics from "On the Amazon" (Greatrex Newman-
Vivian Ellis) © 1928 Harms, Inc. Copyright renewed. All rights
reserved. Used by permission of Warner Bros. Music.

Excerpt from *Sons of the Sun* by Marcel Homet. Neville Spearman,
London, 1963. Reprinted by permission.

I am very grateful to those people who have helped me, both in conversation and in correspondence, with the details of this book. They include: Richard DeCombray, John Green, Laymond Hardy, Mary Sargent Hulme, Steven Hulme, Grover Krantz, Robert Morgan, John Napier, Pino and Renée Turolla, The Mount Everest Foundation, and, most especially, Robert Ferro.

Michael Grumley

FOR MY MOTHER AND FATHER

Photographs

Following page 58

1 Frame from Patterson film of 1967 showing female Sasquatch near Bluff Creek, California. (*Roger Patterson, 1967*)

2 Collection of casts of Sasquatch footprints belonging to John Green, October 1973. The cast he is holding is a duplicate of a cast made by Roger Patterson in 1967 in northern California; it measures seventeen inches long by seven inches wide.

3 Giant footprint (13″×18″) photographed by Eric Shipton on the Menlung Glacier in 1951. *Royal Geographical Society*

4 Tracks in the snow near Mount Adams, Washington, 1969. (*Robert Morgan*)

5 Print from Skamania Country Road, Skamania, Wash., made during National Wildlife Expedition, 1970. (*Allan Facemire*)

Following page 82

6 Skamania Country print cast, held by Robert Morgan. (*Eve Phillips*)

ix

7   Jim Butler with (*left*) North American print and (*right*) Eric Shipton's Himalayan print. (*Robert Morgan*)

8   Jim Butler's foot in The Dalles (Oregon) Sheriff's office cast. (*Robert Morgan*)

9   Pueblo rock paintings, Abo settlement, Abo, New Mexico. (*Robert Morgan*)

10   5″×7″ Jadeite amulet found by Pino Turollo in 1970 expedition to Guacamayo Range during Mono Grande attack. (*Pino Turollo*)

11   Killed in 1924 in Venezuela by De Loys, a giant ape. (*François de Loys*)

There were giants in the earth in those days; and also after that, when the sons of God came in unto the daughters of men, and they bare children to them, the same became mighty men which were of old, men of renown.

Genesis 6:4

Picture a familiar scene.

A blond young woman wearing a pale shift huddles in the niche of a high rock cliff. On her face is a look of terror, and she screams as no one has ever screamed, before or since.

The scream is one of fear and alarm, and it goes on and on and on throughout the length of the film of which this scene is only one frame. The scream is to a large degree what the film is all about—fear and alarm, hysteria and panic, but also, every so often at its lowest register, something akin to fascinated animal attraction. This fascination is not ever allowed to break through to passion, and the young woman behind the scream does a good bit of terrified swooning. But her small breasts do heave beneath her shift, and as she thrusts herself forward from her niche between swoons, only to recoil backward a moment later, there is in her movements the undertone of an involuntary dance of approach-avoidance. This subconscious tango—cringe and squeak, thrust and shriek—is one triggered by an amazing superhuman tableau being staged before her, and the few dark notes mixed in with her pristine soprano are inspired by its main actor.

For he who prompts this ambiguous response is the biggest hunk of hominid flesh that has ever trod the cinematic boards: King Kong. At the moment of Fay Wray's present frame of distress, he is wrestling with a python as long as a Metroliner, grimacing and bellowing and flexing his enormous hairy muscles all the while. He is magnificent and deadly, attractive for his incredible girth and strength, and yet repellent because of these same qualities. He is the most gigantic giant of them all, an anthropoid in the grand tradition. He is the myth and ancient legend of a hundred cultures magnified to their highest, their tallest, their heaviest degree. He is as much an archetype in the universal subconscious as the enormous serpent with which he wrestles; Freudians and Jungians alike claim him as a symbol. He is dark, evil, cunning, and ferocious—all that he is represents the principle of brainless brawn. Mostly, he is big.

Anyone or anything that can scale the Empire State Building and swat down airplanes as if they were no more than pesky mosquitoes has got to be a hero, even though his conduct and character seem hugely misanthropic. And Kong is a hero, no matter what his villainy otherwise consists of. Such bigness is power, and all the world responds to power, finding it admirable no matter which side of the human/animal line its possessor happens to be on.

This is a book about bigness and power, and about the human/animal line. It concerns the present and it concerns the last ten million years, and it has to do as well with the screams of all the Fay Wrays past and present. It is about the existence in the world today of three disparate but not dissimilar strains of such giants. It is about King Kong himself, plucked from the screen, whittled down to a third of his screen size, and made quite verifiably real. It is a book about this Kong as the brother of us all.

One

During the last half-century, anthropologists have made a number of discoveries that are of special interest to those involved with the business of tracking down giants. One of the most significant of these was made by a Dutch geologist and paleontologist, Ralph von Koenigswald. While on a study trip to Hong Kong in 1934, Dr. von Koenigswald came across a number of large teeth lying about in jars in a Chinese chemist's shop. There was one tooth in particular that caught his attention, as it most surely would have caught the eye of even the most unknowledgeable tourist. This was a third lower molar. It was twice as large as the corresponding tooth in the mouth of an adult male gorilla. The volume of the tooth, compared to the volume of a man's tooth, was five or six times as great.

With this first discovery in hand, Von Koenigswald was able to reconstruct a hypothetical primate which would have stood somewhere between eleven and thirteen feet tall. Further investigation over the next five years turned up two more of these giant molars, which the Chinese called dragons' teeth and which were invariably found second-hand, by someone's grandfather out in someone else's backyard. From these first three "dragons' teeth" Dr. von Koenigswald reconstructed a creature that he called *Gigantopithecus*, which today has over one thousand addi-

1

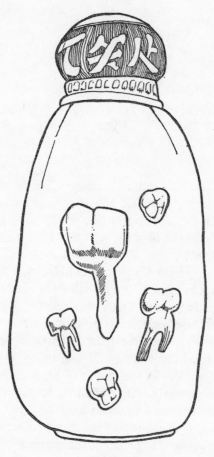

tional teeth to its credit, due to the efforts of the Chinese Academia Sinica in and around Liucheng, and various other bodies. Unfortunately for exploration at the time, Dr. von Koenigswald's activities were curtailed when he was imprisoned by the Japanese, as an enemy alien, during the Japanese occupation of the East Indies during World War II. In 1944 the late Franz Weidenreich assessed the importance of the imprisoned doctor's work and declared that *Gigantopithecus* was not only a giant ape but actually more of a giant man, and stood as one of modern man's ancestors.

In the spring of 1956, in the same Kwangsi province of South China where the majority of teeth have been pulled from the soil, a jawbone belonging to *Gigantopithecus* was found in a phosphate cave. A number of things have been deduced from it by its discoverer, paleontologist Dr. Pei Wen-chung. First, the evidence of the jaw, a lower mandible, suggests that it was chewing a mixed diet of vegetables and grains and an occasional bit of meat, somewhere between 400,000 and 600,000 years ago. This would place it smack in the middle of the Middle Pleistocene. Dr. Pei further suggests from the proportions of the jaw and its teeth that the primate to whom it belonged was at least twelve feet tall and was an ape, after all, and not a man.

Now, the difference between being called a man and being called an ape is a very interesting difference indeed. There are those scientists who would speak of all primates who occurred before *Homo sapiens,* or at the outside, before *Homo erectus,* as being apes, not men. That interpretation of hominid semantics would mean that there were, before approximately 250,000 years ago, no men on the earth at all. But the recent discoveries of Richard Leakey, son of the venerable Mary Leakey and Louis S. B. Leakey, are pushing back quite convincingly the existence of man another 2.6 million years.

Clearly, the difference is an important one, but, unfortunately, one man's man is another man's monkey, and the line of evolution is often knotted with semantic tangles. The new discoveries by Leakey *fils* do tend to shuffle the existing anthropological cards of Leakeys *père* and *mère,* however, in ways that have much to do, not with linear progressions, but with concentric circles of existence. Richard Leakey's recent find at Lake Rudolf in Kenya indicates

3

that there was an amazingly modern looking anthropoid walking around long centuries before he had been supposed to be—and that this early man was not the ancestor of the line that led to the later, beetle-browed *Homo erectus,* but was instead an already refined alternate route on the evolutionary map. He had shot out at a tangent from the more cautious line of development and managed to refine his brow and chin thousands of years before such specialization was the primate vogue.

To clarify the position of these various apes, men, and giants, let us point out that scientific opinion today recognizes, albeit grudgingly, that there is not merely one line of human development, one pattern exclusive of parallel development and improvisation. The evolution tree has many branches—often, in fact, resembling more a bramble thicket than a simple elm or ash—and from these twisting branches any number of discovered and undiscovered species may swing.

Nature published in August 1972 an interesting article advancing the proposition that members of the species that we term *Homo erectus* were alive and well and living in the Kow Swamp region of Australia less than ten thousand years ago. This does not mean that the last descendants of this species, generally thought to have vanished 200,000 years before, were still clumping about. Rather, it means— according to Dr. A. G. Thorne and Dr. P. G. Macumber— that the species father outlasted the species son, with forty well-preserved skeletons bearing out the proposition. It's quite a leap of time and logic to think of certain species hanging on, through ice age after ice age, even as the continents were being pulled apart by the reversing magnetic poles, and as the oceans were breaking on shores that had never been shores before. To think of such species longev-

4

ity is to think of a hardy group of fellows indeed, and in many cases the bigger these fellows were, the better.

Body bulk can be either a plus or a minus when it comes to long-term survival, depending on climate variables and existing predators; often the need for warmth and the need for mobility are at opposite ends of the adaptive scale. But brain bulk would seem to be a thing devoutly to be wished for on the road to ongoing adaptation and survival. In terms of cranial capacity among the primates, there is considerable difference among species and among genera. The earliest australopithecine with which we are familiar had a brain capacity of five hundred cubic centimeters, while the latest Leakey skull, with its many advanced cranial features, has something like a capacity for eight hundred. The modern African gorilla has room for around six hundred, and man today, the paragon of animals, has fourteen hundred. There are as yet no universally accepted figures for the gigantopithecine brain capacity—it is tempting but not reasonable to deduce cranial pith from the lower jawbone of one who was so exceedingly long in the tooth.

There is, however, between these different measurements, a considerable amount of room for variation. And weight alone, the pure heft of the cerebral mass, is not any true determinant of intelligence; a man's brain today, for instance, comprises only one forty-sixth of his total weight, whereas that of a spider monkey makes up roughly one sixteenth of his.

Even with our current computerized calipers and technological protractors brought to bear on brain size, weight, and over-all design, it is still quite impossible to determine what, in direct proportion to one another, these various brains actually do. The most we can claim is that we understand, up to a hypothetical point, how it is that man thinks,

5

but never can we claim that we know what he thinks, nor why.

The correlation between brain size and brain power for any two thinking primates of the same species—for, say, a Fischer and a Spassky—can only be a correlation without measurable correlatives, a graph on which all the lines are written in invisible ink. The nouns are all we know so far; we haven't any idea of what the verbs are all about, and it is the verbs of thought that are all-important. It is the sizzle of the brain pan that signifies, and not the size alone.

And so it is with our giants. It would seem that there is no way of knowing whether *Gigantopithecus* and his immediate relatives were hugely cunning thinkers or merely colossal dummies. But whatever their neural processes, we share a primal kinship with them; the holes in our heads are not so very different from the holes in theirs. What remains to be seen and proved is whether or not *Homo sapiens* is the only *Homo* or near *Homo* who experienced and developed the patterns and abstractions that make up what he today refers to as "thought." It is just possible that along the line there may have been other primate flashes in the cerebral pan.

In the last year of the reign of Queen Victoria, there appeared for the first time in London and New York reports of a strange form of life which was to be found high in the Himalayas. Just as the new century was about to unfold, with all its subsequent talk of apes and angels and its crosses of gold and war, Major L. A. Waddell published in England his book *Among the Himalayas*. Perhaps the most interesting part of the account he gives of life in that most inaccessible of mountain ranges was his discovery of a number of large footprints in the snow of northeast Sikkim. He writes, "These were alleged to be the trail of the hairy wild men who are believed to live amongst the eternal snows."

Woven through the folklore fabric of the entire Himalayan Range, from the Karakoram to northern Burma, in Nepal, Tibet, Bhutan, and Assam and in the tiny but glittering kingdom of Sikkim, there had always been stories and legends that concerned enormous creatures, half-man and half-beast. But these stories, up until the time of Major Waddell's publication, had seldom found their way to Western ears; they were the tales of the uneducated, the nonscientific. And they came from time out of mind.

Later, in 1921, the first attempt to climb the northern face of Mount Everest was made. Colonel Howard-Bury and the rest of his party of climbers and bearers were fortunate enough to see great dark spots moving across the snow valleys of the Lhapta-la Pass. The Tibetan porters were in no doubt as to the nature of the distant beasts; they were unquestionably members of the race they called *metoh-kangmi.* The translation of this phrase into English has given us the unfortunate term "abominable snowman." Such sightings were not new to the natives, and, when later in its ascent, at an elevation of 23,000 feet, the party overtook a path of enormous footprints, the bearers were not in the least surprised by either their size or their configuration. The Western mind was not so familiar with snowman prints as the Eastern, however, and the official report of the party called these enormous tracks the work of some large stray quadruped—even though they came in twos and not fours—and decided that they were left by a wolf.

7

Since then, over the space of the last fifty years, such tracks have been credited to the Nepal langur, the snow leopard, the Tibetan red bear, Roxellana's snub-nosed langur (or snow monkey), and even to the feet of the various monks and pilgrims who make their homes in mountain lamaseries. Each of these possibilities has been quite thoroughly explored by any number of reputable anthropologists and zoologists, and over each little bundle of feline or ursine explanation, there has come a new set of prints that disproves it. In fact, the sightings of tracks have increased to such a degree that each new find, complete with plaster casts and still photographs, is greeted by the scientific community with no more surprise than that which Howard-Bury's bearers displayed. Expeditions, organized by such strange academic bedfellows as Prince Peter of Greece and the late Texas oil magnate Tom Slick, have come up with a sizable amount of data on the habits and appearance of the elusive "snowman," yet never with the furry gentleman himself.

But the creature is alive enough in theory to have already been divided into three Himalayan types, and to have been given, in Bernard Heuvelmans' excellent book, *On the Track of Unknown Animals,* its own Latin name, *Dinanthropoides nivalis.* Heuvelmans describes the animal thus:

> We have reason to believe that it is a large biped anthropoid ape, from five to eight feet tall, according to its age, sex, or geographic race, which lives in the rocky area at the limit of the plant line on the slopes of the whole Himalayan Range. It has plantigrade feet, and the very conspicuous big toe, unlike that of most monkeys, is not opposable to the other toes. It walks with its body leaning slightly forward; its arms are fairly long and reach down to its knees. It has a flat face, a high forehead, and the top of its skull is shaped

8

like the nose of a shell; its prognathism is slight, but its thick jaws have developed considerably in height, hence the disproportionate size of its molars. To this outsize masticatory apparatus are connected very powerful jaw muscles. On the cranium there is a sagittal crest which is revealed by a thickening of the scalp, in the adult male, at least, and the presence of upstanding hair. It is covered with thick fur, which in the smaller specimens varies from fawn to dark chestnut in different places with foxy-red glints, but the face, chest, and lower legs are much less hairy. In the larger specimens the hair is an even darker brown or almost black. It appears to be omnivorous; roots, bamboo shoots, fruit, lizards, insects, birds, small rodents, and occasionally larger prey like yaks are all grist to its mill in such barren country. Its cerebral capacity should be about equal to, or even greater than, man's.

The tracks of the *Dinanthropoides nivalis,* or more simply, the Yeti, crisscross the Himalayan Range in increasingly evident numbers. But it is not only in this Asian mountain range that the appearance of a "large biped anthropoid ape" is stirring up the snow. There are two other mountain ranges where the folklore of the natives tells of such a creature, and where its physical description closely parallels that of the more familiar Tibetan giant.

The second area where our giant biped has crept out of obscurity and shown himself to the modern world is an area where the cataloging of known forms of wildlife is as much a challenge as the discovery of those as yet unknown. The theatre of discovery is the Green Continent, South America, and the stage on which our second character performs is a proscenium which stretches a length of over four thousand miles—the Andes Range.

The first reports that come out of South America come from the northernmost tip of the Andes. A member of a Spanish gold-hunting expedition in the late seventeenth century wrote back to Sevilla that his group had fought with and killed fourteen giant beasts, in a section of jungle near the Colombia-Panama border. Subsequent stories of "giant orang-utans" and "man-apes" drifted back to Europe from time to time, as the tropical forests and mountain slopes were crossed by Spaniards, Portuguese, and Englishmen intent on pulling precious minerals out of the earth and sticky resins out of the bush. The names of the beast they encountered were numerous—known as the *mapinguary, capelobo,* or *pelobo* in the jungle vocabularies of the south, across the Amazon delta through Brazil, or as the *di-di* or *Mono Grande* in the northern countries of Ecuador, Colombia, and Venezuela.

Richard Oglesby Marsh writes of an American prospector named Shea, who in 1920 arrived in Panama with a tale he'd brought up from the Andes, a tale centering on his encounter with an animal "six feet tall, which walked erect, weighed possibly three hundred pounds, and was covered with long black hair." Shea's encounter amounted to him being surprised by the animal standing upright before him on a mountain ridge, chattering at him angrily, and of him pulling out his revolver and promptly shooting it through the head. Examining the dead animal where it lay before him, Shea noticed that the big toes on its feet were parallel with the other toes, as in a human being, and not opposable and thumb-like as they would have been on the foot of any monkey or ape.

In the same year in another part of the South American

10

forest, another man was meeting up with his own furious primate. Swiss geologist François de Loys was returning with his expedition from the Sierra de Peija in the northern tip of Colombia and was making his way along the Sierra de Unturan, near the Brazilian border of Venezuela. Not far from a small river, he came upon a pair of screaming "monkeys" who were so beside themselves with the rage of being discovered they filled their hands with their own excrement and threw it at the party of intruders. In response, the men shot the nearest animal, with the other escaping into the brush. The corpse of what turned out to be the female of the pair was propped up on a fuel crate and photographed by the geologist; the photograph of the dead female survives today.

This photograph shows an animal not much over five feet tall, yet in a hemisphere where the largest monkeys are never more than three foot seven, this specimen does seem to be a rather curious savage. De Loys counted thirty-two teeth in the animal's mouth, which is a surprising number for any New World primate; platyrrhinians, or New World monkeys, without exception have thirty-six teeth in their heads, while catarrhinians, or Old World monkeys, have thirty-two. (This is not just a question of existing canines and molars, with the possibility of decay or other loss; it concerns jaw structure and tooth placement just as much as tooth number.) This means that the angry animal had no more business in Venezuela than did the Swiss scientist; both belonged, ethnographically speaking, back on the other side of the water. That the outsize monkey was larger than most Old World varieties, and all New World ones as well, raises questions about the validity of so separating genera by nose-and-tooth generalities. Perhaps the definition of the difference between catarrhine and platyrrhine is presumptuous in the extreme, and has much less to do with the real flora and fauna of the earth than with those floating islands, called continents, on which they currently find themselves.

In this context, let us consider the third main character in this book—a bit of folk legend made real by seven minutes of sixteen-millimetre film. The animal involved in the filming is called Bigfoot in the Pacific Northwestern United States, and, more than the Yeti of the Himalayas or the Mono Grande of the Andes, this creature is clearly the anthropoid of the hour.

Close to eight feet tall and weighing nearly a thousand pounds, leaving tracks that measure fourteen and a half inches in length, the animal is recorded on film taken by the late Roger Patterson in October 1967. Patterson and his companion, Bob Gimlin, were out horseback riding in the Bluff Creek region of Northern California, checking on reports that there was someone or something leaving giant footprints in the region, when that something strode out of the bush and moved directly in front of them across the trail. It made no threatening movements, but looked the two men full in the face as it walked. The horses were sufficiently unnerved to shy backward away from the animal, in the process throwing Patterson and Gimlin to the ground. Before it disappeared back into the scrub pine and underbrush, Patterson was able to grab his movie camera and focus from where he lay on the retreating animal. The result is a film showing a huge hairy upright creature, which, like De Loys's monkey, appears to be female—its furry breasts are prodigious—and which gives positive substance to the giant ape stories that have been trickling out of the Pacific Northwest area for the last hundred years. Since 1958, when various sightings and print reports breathed life into the old Indian and prospector legends, there have been recorded some 750 sightings of either the

12

animal itself or its tracks. These reports, along with Patterson's film, make up a considerable provenance.

These three giants, of whom more detailed mention will follow, are the three characters whose activities make up the far-ranging Noh play—with all their stylized gestures being today acted out on three continents—that is the current subject of considerable speculation and scientific interest in both the popular and the academic press. The first question that must be asked is How can they possibly exist at all? Do they really form an ensemble company working from the same ancient script, or is the only thing that unites them the whim of some offstage costume mistress? The answer to this question is not an easy one, but it deserves all the detail we can muster and all the imaginative curiosity we can bring to bear. There are clues in these furry giants' existence that might help solve some of the mysteries of our own; by presenting the facts of their lives it is more than possible that we will come up with some new theories about ourselves.

But we are incidental. Modern man is, after all, a rather runty specimen when compared to the dimensions of the rest of the evolutionary zoo; for size and perseverance we had best look elsewhere. All giants are not monsters, and, certainly, all monsters are not giants.

It is all a question of extremes.

Two

Before we begin heaping up and sifting through the mound of outsize anthropological evidence, let us leave for a moment the rooftops of the world and take a quick trip down into the cellar. Down in the deeps of the oceans there exist specialized and improbable creatures who, when it comes to adaptive innovation, put their land-roving cousins to shame. The elusive anthropoids of the Himalayas, South America, and the Pacific Northwest seem as natural and straightforward in appearance as a herd of Holsteins at pasture, compared to these gilled wonders.

The order of Lophiiforms, which includes both the anglerfish and the frogfish, is an order notable mainly for the lengths its members have gone in the name of specialization. The frogfish has had the physiological cunning, over a period of countless generations swimming at deeper and deeper levels, to exchange what was originally a pair of perfectly forthright pectoral fins for a pair of grasping scaly hands. Able to provide locomotion, as well as reach out and grasp the advantage over its less efficiently equipped competitors, these appendages with their discrete fingers neatly illustrate one species' method of coping with the problem of heavy aquatic traffic and light food supply. Whatever it was that inspired the frogfish to so consolidate his fishy genes, it bespeaks enormous pluck and also a certain modicum of good taste. (Diving lower and lower in the inky oceans, it perhaps seemed to the first innovator that, in the words of Parmenides, the way up was the way down, and the new hand-to-mouth feeding method was the result of suddenly finding its fins on the ground.) This tasteful bent is one remarkably absent in the adaptive conduct of the frogfish's next of kin, the anglerfish, who has gone out on a rather terrifying adaptive limb.

The anglerfish's method is nothing if not extreme. When we are not referring to them as anglerfish, we call them instead sea devils, thereby showing what we mammals think of their excessively piscean pragmatism. The first of their specialized devices is not so interesting as the second; it is a device used by other species, both vertebrate and in-, and is a fleshy lure that is used by the anglerfish as bait to attract other smaller and eminently edible fishes. (This lure, or illicium, is used by the frogfish as well, but its armlike pectoral fins are really what *its* adaptation is all about.) The second device is one that belongs to the anglerfish alone; it represents what we primates would call a very unhealthy relationship. Yet it works. Roughly an inch above the anglerfish lure, there dangles what seems to be another lure still in the process of becoming; the proper lure has a luminous tip and a tendency to shimmer, while this second growth is rather dowdy in comparison (and may occur anywhere on the fish's body, but most often does in the illicium region). This second growth, if a particular fish is eleven to twelve inches long, will be from two to two and a half inches in length. It is not luminous and is not, strictly speaking, bait. It is, merely, the male of the species.

That grizzled Nioka, the lady anglerfish, has taken to wearing, in the name of survival, her husband like a hat. And the male, with his specialized hooklike teeth and his

15

body size only one sixth of his beloved's, seems not at all dissatisfied with the arrangement. Early on in their conjugal life, he anchors himself to her body, and very gradually he becomes part of her, in the most literal sense. His mouth slowly stretches apart, and week by week, hour by hour, all the parts of the mouth and the digestive organs are taken in by her; the organs grow together and the two circulatory systems merge. Finally, little, if anything, is left of the male, except for a few fin remnants and two proportionately enormous testes.

He is a feather in her cap, obliging to the end. It does seem a strange way to get through life, but there is always the chance that the male anglerfish knows something we landed chauvinists don't. Why he chooses, generation after generation, to stay locked in this thoroughly self-effacing embrace is a problem for more advanced ethologists to ponder. Perhaps he replaces that fin which has become a lure, or perhaps he has not, as it would appear, abandoned all say in the direction their life will take, embedded as he is as a kind of fleshy rudder. It is difficult to know.

Less difficult to understand, but equally intriguing, is the behaviour of that sluggish, yet lethal, inhabitant of the Red Sea and western Pacific, the stonefish. Besides looking as unpalatable as he could possibly be, the stonefish carries on its back thirteen of the deadliest spines in all the animal world. The stonefish has always been acknowledged as the most poisonous of all its teleost family, even more than the innocuously named but venomous rabbitfish, who also boasts thirteen poison barbs along its back.

(It is tempting, although outrageously facile, to surmise that when our cultural elders were organizing human knowledge and experience into the cryptic letters and cards of the Hebrew alphabet and Kabbalah, thousands of years ago beside the stonefish's home seas, they were influenced by the number of spines on his back. The thirteenth letter and the thirteenth Tarot card were thereby representative of Death, and the number thirteen a reminder of caution and dread. [Fish, unfortunately for this theory, care nothing for numerology, and conceal three more poison spines in their anal stonefish fin and two in their pelvic fins. This adds up to eighteen, not thirteen, but, luckily for Kabbalists with fishy affinities, the eighteenth Tarot card is the Moon, and the Moon's pictorial translation has a decidedly lethal-looking crayfish emerging from the water at the bottom of the pictogram, claws at the ready.])

Whatever the stonefish's spines mean, or have been translated to mean in numerology or Tarot, they mean out-and-out spine-tingling success in terms of ocean adaptation and survivability. The stonefish spends his days well camouflaged as a bit of warty rock, a grisly non-aggressor. Yet, when and if he is detected, he still has the capacity to kill within a matter of minutes any eagle-eyed shark or feisty barracuda hungry enough to nose him out of his shelter. For such an ugly little fellow, he manages quite well, and it may all be said to be a question of self-defense.

Spines and fins do amazing things in the fish world, having provided all manner of shocks, lures, lights, rudders, lungs, and sexual come-ons ever since the first fish-like animals, or ostracoderms, appeared in the primal waters 450 million years ago. These jawless ancestors of all modern fishes were joined in Silurian times by jaw-bearing placoderms, and it has been one long labyrinthine road of specialization ever since.

There have, however, been reactionary factions in the politics of evolution, and the most visible right-winger of them all these days is a breed of fish which has followed his own lights since the Middle Paleozoic, undaunted by the progress of his libertine in-laws, true to the fossilized Philistines of his youth—the redoubtable coelacanth. Until 1938 thought to be only a fossil of his former self, the coelacanth in that year turned up alive off the coast of Madagascar, with his Upper Devonian credentials still intact, differing from the hoariest of his ancestors only in his slightly smaller swim bladder. Swimming blandly through time and the sea from the Middle Paleozoic on into the Mesozoic, and through all the evolutionary upheavals of the Cenozoic era—with neither a discernible past nor a seemingly alterable future—this crusty specimen is older than the breaking up of the continents, older even than the separation of the seas themselves.

The present-day coelacanth, *Latimeria chalumnae*, named for the Chalumna River in southeast Africa from

18

whose depths he was lately plucked, appears with every catch that is brought in to be less and less of a one-of-a-kind curiosity. Since 1938, a number of other specimens have been piling up in the zoological hold, all taken off the Comoro Archipelago northwest of Madagascar.

Described by N. B. Marshall,

> *Latimeria* is very like its fossil relatives, which are unknown beyond the Cretaceous period. The heavily built body, which is dull grey-blue in colour with irregular white spots, is covered with cosmoid scales. There is a pair of nostrils on either side of the snout. Between the nasal organs is a fibrous sac that is well equipped with nerves.

Rudimentary but efficient is the coelacanth, slightly smug in its early precision—with a proper nose, a body five feet long, and a weight of 180 pounds.

> The rostral organ, as it is called, is evidently some kind of sensory device. . . . As the male has no intromittent organ, fertilization (of the female's eggs) may well be external. If so, the incubation period is likely to be long and the young must hatch at an advanced stage. *Latimeria* grows to a length of over five feet. . . . By and large, it exists by preying on other fishes.

Now, 180 pounds is a good size for any fish that has swum through so much time and water; small fishes, like small mammals, have more places to hide, but are also easier prey. And hatching at an advanced stage is certainly prudent. A nose so early on is an equally sophisticated touch; it may well have been, even at that point, more sophisticated and specialized than our own human noses today. (As Virginia Woolf, in *Flush* remarks, "The human nose is practically non-existent. The greatest poets in the

19

world have smelt nothing but roses on the one hand, and dung on the other.") It seems a pity that after all these millions of years since it came about, the nose of the coelacanth should survive to sniff only at Australian oil slicks, French atomic-bomb installations, and at the only known slum-thriver, the sea viper in its bower. Still, judging from its endurance record, *Latimeria* has wriggled through enough earthly upheavals to weather another holocaust or two, and presumably will be able to outlast the current favourite, the viper, when the waters once again clear and the sea is once again the sea. When man is only a bit of dust and algae again, I imagine the coelacanth will not even bother with a gurgling I-told-you-so sigh, but will simply follow his infallible nose into the watery future. The sea is, after all, so very large.

It is mainly along the rivers of the world, and not in the oceans, that the fish are beginning to get a bit peevish. Walking catfish are leaving the Florida swamps for higher and, we suppose, cleaner ground, and down along the Amazon the piranhas are snapping at bits of Polaroid negative and plastic Pepsi-Cola bottle with increasing fury. On the Amazon, where as Bobby Short fancifully sings,

> The prophylactics prowl, and hypodermics howl,
> While duodenums are lurking in the trees—
> And the jungle swarms with green apostrophes . . .

there abound likewise some very real and fully fleshed dangers as well.

On this unrivaled giantess of rivers, there is one particular catfish, a tiny fellow with as much curiosity as his feline namesake, who makes a habit of swimming into any little orifice that presents itself in the warm waters. The Tupi Indians of Brazil refer to him as the *candiru,* and they like him not at all; it seems those orifices he is most fond of exploring are those of the human body. Once he has wormed his way into one's aural, nasal, or even anal canal, his barbed gills make him all but impossible to remove without surgery. His presence soon becomes exceedingly tiresome to any unlucky host, as may be imagined.

More visible, but certainly as vexatious, is the huge *piraiba* catfish, who at a scaly bulk of 350 pounds has been known to gobble down small children who stray into the River Sea's deeper channels, swallowing them slowly one after the other like bits of hominid caviar on his watery plate. There is also in these same waters a fish called the arapaima swimming cheek by slimy jowl with the voracious piraiba, but whose appetite, happily, eschews humankind. He holds the honour of being the world's largest teleost fish, and may measure twelve feet from the tip of his bony tongue to the end of his languidly thrashing tail. (The world's largest rodent may also be seen disporting himself in South American waters; called the capybara, he is largely aquatic and is found nowhere else but on the Green Continent.) This enormous arapaima, the largest fish of the largest river, has managed to survive nearly intact from Cretaceous times, clinging still to an archaic palatal structure that he shares with only four other species of fish life, out of a possible roster of eighteen thousand teleosts. Near Manaus, the Brazilian trade center on the Amazon which at the turn of the century was the world's rubber capital, this giant fish is called the pirarucu and is thought to be quite delicious when fried or broiled or made into a hefty pirarucu stew.

Such weighty zoological anachronisms abound in the

river and across the surrounding delta, and there are in addition to the visible beasts some interesting fable beasts lurking in the area. One such creature is the legendary curupira. This animal is a curious type who shows up in various Amazon folktales, with various attributes depending on the region of the tale itself. He is often referred to simply as the "hairy man of the forest" and is said to have sparkling green teeth, to be never more than five feet tall, and to mimic human speech. He ranges over the entire delta, from the Rio Juruá to the Rio Xingu, and is generally regarded as a jungle prankster, more on the order of a leprechaun than a towering troll. He is no giant, and it is quite likely that he doesn't exist at all. Yet there is some connection between the names of the mammoth pirarucu and this manlike curupira—the old Indian names indicate an affinity that is unclear in the actual Indian tales, an affinity that has more to do with an earlier association of ideas, perhaps, than the one now in legendary vogue. Frieherr von Humboldt was intrigued with the similarity during his Amazon trips from 1799 to 1804 and wrote about it at some length. Later, in the 1870s, there was a spate of interest in the curupira's provenance, and two Brazilian monographs still exist from that time. The animal, if he does, after all, exist, appears to be particularly Brazilian—in his more fanciful moments he is able to rise full-bodied and restored from his own decaying skeleton, and in his more mundane existence is fond of banging on tree trunks, and of screaming a lot. But it is his name that seems a nattering footnote to a body of language now forgotten or obscured; unfortunately for serious primate-hunters, this fuzzy obscurity has in the past covered any real tracks with his own whimsical haze. The jolly green fellow, the curupira, has become the Judas ape for all other man-of-the-woods stories and sightings, even when the sub-

22

ject of such tales is eight feet tall and resembles the improbable curupira only in the hair of their mutual coat. To dismiss the curupira because of his odd habits—cutting out his heart to exchange for an Indian's, possessing feet that leave only backward-pointing tracks, and being able to appear and disappear at will—is an easy way of dismissing at the same time all reports of as yet unclassified hominids who inhabit the same area or, at least, the same continent. It is indeed convenient to say that if the curupira is only a myth, so then is the mapinguary, the di-di, the capelobo—all those taller, shriller types who have been glimpsed and described by outsiders and who ought to stand a good chance of being glimpsed again and again, even, perhaps one day, by some degree-toting anthropologist.

Never mind. The curupira will have to remain a forest enigma scampering forever backward through the soggy Amazon delta. Let us allow him to frolic there as he will, riding through the jungle on the plated back of some equally improbable megathere, if he likes. He can stay with the mammoth fishes who are, as likely as not, quite happy to have him around.

We, meanwhile, will endeavour to proceed to the flesh-and-blood big boys, the game that this book actually seeks —those South American primates who may turn out to be this legendary monkey's corporeal and substantial uncles.

It is an easy thing to stretch the truth in someone else's language. Somehow, exaggeration and misinterpretation come more easily when the tongue one is using is not one's own and where the metaphors of another culture too readily take the place of facts. What the Sherpa bearer is describing to the British journalist in halting detail; what the French Canadian trapper confides to the reporter from the Ketchum *Herald;* what the Waika guide is soberly re-

lating to some Argentinian Lois Lane—these are not always the same descriptions that each might give to his own cultural intimates.

But in the case of the language being thoroughly understood by the teller and listener alike, and of the Immelmann twists and turns of the dialect being followed by both, there is considerably less chance of getting lost in the linguistic clouds.

Pino Turolla is an archaeologist with a remarkably well educated ear. Above and beyond his Italian talent for separating what people mean from what people say, Count Turolla has an intuitive talent for understanding what people mean they mean, as well. After a decade of exploring the remote jungles and mountains of South America, he has an easy familiarity with a good many Indian dialects, most particularly those of the Upper Amazon basin. In situations where a simple gesture or phrase will mean completely different things, depending on context and emphasis, to different tribes, he has been forced to master the nuances for the sake of survival alone.

Turolla's investigations have been mainly concerned with the existence on the Upper Amazon range of a sophisticated stone-carving culture, a culture that thrived long before the clay age and long before the current estimates of 3000 B.C. The stone artifacts that he has collected as evidence of this culture are amazing in themselves; his display cases are crammed with elephants, camels, and all manner of beasts not generally thought of as South American. To see this stone menagerie is to realize how far back the concept of culture really stretches; the *pre* in pre-Columbian covers such a multitude of sins of anthropological omission.

It was while tracking down such artifacts in the caves of Venezuela that Count Turolla heard reports of the legendary Mono Grande. In April 1968 Turolla's Zapes Indian guide, Antonio, described his own exposure to the animal and the tragic death of his own son. Antonio had gone with his two sons to the Forbidden Range of Pacanaima, and as the three men approached the savannah, they met up with three enormous creatures. Described by Antonio as huge lumbering beasts with smallish heads and extremely long arms, the three set upon Antonio and his sons with clubs. During this attack, the younger son was killed.

Six months after hearing this story, Turolla returned to Venezuela and convinced Antonio to guide him to the Forbidden Range where the attack had occurred. This time, though, there was no confrontation; close to the Pacanaima ruins the animals circled the party through the bush, but did not attack. Instead they set up a howl which Turolla describes as being in volume like the roar of a lion, but in pitch higher and more shrill. The three Indians in the party became nervous, and then alarmed, as this howl continued, and after a few moments they turned and ran back along the trail. The howl continued, a terrifying sound that seemed to come from all around them at once, drowning out all other noises, rising again and again as Turolla advanced toward the savannah in the distance.

But nothing happened. After a stillness between the howls, the animals—this time there seemed to be only two —loped off through the foliage, and the one image that Turolla retains is of their enormous shadow moving across the boulders before him in the dusk. No longer howling, they soon became a blur against the rocks; the impression he had was of two hairy apelike creatures, well over six feet tall, who ran and leapt in an erect posture. No emerald-green teeth. No feet facing backward as they leapt. Just two perfectly straightforward giant apes, howling, as Turolla describes it, like the devil himself. He estimates their height as being between six and eight feet, but stresses that he was over fifty feet away from where they were silhouetted against the rock. There were no other Indians and no prospectors in the area at the time. After the animals had disappeared, he shot his rifle into the air to call back the men who'd run off, and the group made their way to La Esmeralda. There, white settlers assured him that the "Monos" were not uncommon. One prospector recounted how his small mule had been carried off by one. The animal, when it was found several kilometres away, had been ripped open and torn apart.

Count Turolla had brought with him in his wallet a copy of the photograph of the large monkey, or ape, shot by François de Loys, and found that those he talked to who had seen the local animals identified them closely with the animal in the photo.

There are other stories.

During his South American expedition in 1595 and 1596, Sir Walter Raleigh reported that although he came face to face with none of the animals, he was convinced of their existence by the many tales of those people who had encountered them.

In 1796, Bernard Heuvelmans reports, Dr. Edward Bancroft brought back from Guiana a tale of an "orang-utan"

> much larger than either the African [the chimpanzee] or Oriental, if the accounts of the natives may be relied on. . . . They are represented by the Indians as being near five feet in height, maintaining an erect posture, and having a human form, thickly covered with short black hair; but I suspect that their height has been augmented by the fear of the Indians who greatly dread them.

(Five feet, again, is not so very large, except when measured against the height of all other South American monkeys, who stretch no further skyward than three and a half feet. And there are no known apes in all of South

America—only the many species of platyrrhinian, or "broad-nosed" monkey.)

The attitudes of the Indians and prospectors are markedly different when it comes to their primate neighbors, real and imagined. Where the attitude toward the impish but ephemeral curupira is one of amused and speculative tolerance, the opinions concerning the larger, more selective, and decidedly malevolent giant carnivores are ones grounded in fear and apprehension.

Heuvelmans again reports:

Today, there is no lack of rumours about fantastic man-shaped monsters in the Green Continent, much of which is still little known. Frank W. Lane writes:

Gold prospectors, working on the River Araguaya, which flows near the Matto Grosso, have heard roars coming from the depths of the virgin forests. Cattle have been found dead and, every time, their tongues have been wrenched out by the roots. Two prospectors have seen footprints in soft sand by the river, which resembled those of a man, but were twenty-one inches long.

Henry Pitaud told Heuvelmans in 1955 that a similar event had caused a sensation in the Ybytimí region in Paraguay a few years before:

In an *estancia,* about a hundred cattle were found dead without any wounds except that the tongue was torn out and gone. This went on for nearly eight months. Everything returned to normal, then, suddenly, two or three years ago, the same events recurred in the same area but in a different *estancia:* another hundred cattle suffered the same fate.

Again and again one hears of the disembowelment of goats, mules, and cattle, and the carcasses with their missing tongues and entrails, in Paraguay, Brazil, and even Colombia. As to this reported tongue-yanking and organ-scooping, there is always the possibility that these tales are an extension and projection of the local Indian belief that one gains power by securing for himself certain parts of a living body, human or otherwise. Cannibals still dip their needle-sharp arrows in *curare* along the tributaries of the Amazon, and on the island of Maracá scalps are still taken as a means of gaining the power and cunning of their former wearers. Since this is the case, the organ-grinding proclivity of the Monos speaks here more for their human-ness than it would elsewhere in the world, where humanity thinks of itself as more humane, and where its cultural habits draw a finer distinction between body and soul. (Then again it might just be that cow's tongue is un-speakably tasty for the Monos, and liver-of-mule quite a gourmet treat, unexcelled as far as their fuzzy taste buds are concerned. The whole thing may be simply a question of taste, and not symbology.) (South American appetites *are* curious: There is a tiny two-winged insect called the *mouqui,* which according to Marcel Homet bites only be-low one's waist and above one's knee, leaving all other areas of skin untouched. And he bites thoroughly, and he bites often.)

And there is yet another possible explanation that I would here like to put forward—that the giants sighted, heard, or reported in some parts of South America are in-deed men, after all. I do not believe that this is true of those beasts encountered by Turolla in Venezuela, nor is it possi-ble to look at the photo of De Loys's giant monkey and call it a lady. But I do think it is possible that the outsize creatures cutting up the livestock in the Matto Grosso region of Brazil, at least, are neither so hairy nor so simian as has been reported.

The following article appeared in the New York *Post* on

29

February 26, 1973, under the heading "Tall Story from the Jungle":

Rio de Janeiro—A Brazilian anthropologist has apparently made firm contact with a tribe of giant Indians—many of them six feet six inches tall or more—who speak a language totally incomprehensible to outsiders, it was learned here over the weekend.

Claudio Villas Boas radioed from the Matto Grosso jungle of Eastern Brazil that he and his team on Friday watched a ceremony of the Kreen Akrore Indians in which 30 women and 25 children took part—a sign the Indians wanted peace, he said.

He said the ceremony of the naked giants, whose bodies were painted black, lasted for eleven hours along the banks of the Peixoto de Azevedo River and included song, dance, and the exchange of gifts. Boas had made contact with the tribe for a few minutes two weeks ago after more than a year's search, but then the Indians burned their huts and retreated deeper into the jungle. After the ceremony, he said, the Kreen Akrores slipped into the jungle, waving goody-bye with their hands as they moved toward their main village several miles from the River.

The anthropologist said the giants were in good health—unlike other Amazonian tribes almost wiped out by fever and epidemic—and seemed very "happy, lively, and intelligent." He said the Kreen Akrore live on game and forest products, hunted with bow and arrow, and apparently neither knew how to swim nor about dugout canoes despite their nearness to the river.

Boas said the giants appeared to be "moved" when Indians in his expedition answered their chants with chants of their own. The anthropologist's next step was to attempt to get the Kreen Akrore Indians to use medicines against epidemics.

Later, after this article appeared in the press, there were some other bulletins concerning the tribe of giants; the detail I recall most fondly was that, of the gifts offered to them by Villas Boas and his party, the only ones the tribesmen refused to accept were those made of plastic. Everything else was acceptable—wood, straw, paper—but not the twentieth century's most hideous and most efficient invention, plastic.

This Kreen Akrore, let me point out, is an incredibly remote and small tribe, one that the Villas Boas brothers devoted all their physical and mental energies to locating, over a period of years, waiting through the jungle seasons with their Indian staff, hoping for a meeting such as the one here reported. The Kreen Akrore are as atypical as they can be on this continent of the small in stature; there is nothing that can be said about them with certainty as to their customs and history, except that they do exist. But *where* they exist is highly interesting, especially when one compares these reports from the Matto Grosso of ape-giants—dark, hairy beasts who pull out the tongues of their four-footed prey. The River Araguaya, where Frank W. Lane described the giant footprints, the roars, and the missing tongues, is only 250 miles from the Peixoto de Azevedo area where these Indian giants were finally coaxed out of the bush. Painted black, living on game and forest products—it would not be surprising to learn that it is they and not the ape-giants who are responsible for a good many of the jungle horror stories.

If you like, erase the Matto Grosso variety of shaggy giant from the South American roster. He may very well be out there, snickering at his hairless counterparts from behind a convenient liana clump, digesting the last of a tongue-in-cheek omelette, while they dance for eleven hours along the riverbank. But with the recent appearance of these giant Kreen Akrore, his valid image in the mind's eye must necessarily recede.

As with the improbable curupira, however, one must resist the impulse to throw out all the Venezuelan and Ecuadorian giant-babies with the Matto Grosso bath water. There is still a good bit of evidence to examine elsewhere.

(The main danger, I think, in heaping tale upon tale from miner, explorer, and Indian guide is that each successive firsthand report weighs proportionately less heavily on one's scale of interest and belief. There is something remarkably blasé about one's fact-threshold, and it seems the more experiences that are related, the more belief is diminished, rather than strengthened. One may perhaps be overly suspicious of detail; it is rather like listening to ghost stories of an evening, late and firelit. The embers glow in the grate and the wind whistles outside, and yet at a certain point the goose bumps abruptly fade back into the flesh, and one begins to think of how good a cup of hot chocolate would taste. Just as in art, so in life, too much detail dulls the senses, and Dame Credibility dozes off in a warm corner somewhere.

It is not my intention to usher you through a Gothic cathedral of evidence where, at every window and gutter spout, chimeras pout and grimace, and where one traces a particular line of gargoyle development until—suddenly— unhappily—one's mind begins to wander to the tree-shaded avenues of Saint-Germain and one's mental appetite concentrates on a hot dog in Montparnasse.

This being my fear, I hesitate. I want you, the reader, to believe in what I, the writer, believe, but I don't want to drag you so thoroughly through all the thirteen countries of South America—showing you tracks in a mining camp in Bolivia or the lacerated rooftop of a missionary outpost in Peru—that you begin to yawn and wonder why on earth you consented to accompany me in the first place.)

That said, let me give you one more report, again from

Pino Turolla. He has asked me to put it in my own words, and I shall:

In December of 1970 Turolla was involved in an archaeological expedition in the Guacamayo Range, not far from the River Chancis in Ecuador. The territory that he and his South American assistant, Oswaldo, were exploring was neither specifically Ecuadorian nor Colombian, but belonged instead within the province of the Aucas tribe between the two countries. An Indian shaman from the Amazon gave him a lead, describing to him the entrance to a particular cave in which he assured Turolla he would find evidence to support his theory concerning the real beginnings of world culture, and of South American culture specifically. The two men were pursuing this lead, going through a particular canyon on their way to this special cave, following a trail, or *pica,* that had become merely a track through the heavy foliage. At midafternoon the jungle rain beat down on the two men and their horses, and when they at last reached the area of the cave, they dismounted and ate two cans of sardines before continuing. The rain was so heavy that their visibility extended no more than twenty feet in front of them, and they waited, hoping the downpour would diminish before they approached the cave itself.

At 3:30 P.M. they left the horses some one hundred feet below the cave and climbed to the entrance. This entrance was formed in the shape of a trapezoid, with the opening much smaller at the top than at the bottom. Turolla was at the time puzzled by the precision with which the opening seemed to have been carved; it appeared to be not at all a natural shape, but rather man-made. Even with the rain, the daylight extended into the cave for approximately fifty feet, at which point it was absorbed by the blackness of the deeper chambers. The two men used their flashlights to fol-

33

low the main tunnel, which was roughly twenty feet high and fifteen feet wide. The interior of the cave was completely soundless, without even the sound of dripping water that often punctuates the silence of smaller caverns. After walking 150 feet along the tunnel, the two men met up with a smooth rock wall. To the right, however, a new tunnel began, this one much larger and higher than the first one. The two flashlights picked up small pieces of wood and what seemed to be rock along the wall, and on the ground in front of them the light showed the tracks of something being dragged along, something heavy which had left a deep impression in the dirt as if it had been carried for a distance and then let rest, and then carried again. The men now became conscious of a strong animal odour in the air, a musky smell that emanated from the walls on either side of them. They had proceeded in silence up to this point, but now Oswaldo began speaking softly, as if to himself, not articulating words but only sounds, in his apprehension not really conscious that he was speaking aloud. Turolla led the way along the right wall of the tunnel, now and again putting his hand to the wall, and the two continued for another 250 feet. The only sound was that of Oswaldo's incoherent mumbling. Then, before them, the tunnel opened up into a dome so large their flashlight beams fell short of its ceiling. The tunnel turned left at a right angle and led off into the blackness. Here the two men lighted cigarettes and quietly began describing the cave to one another, trying to put it into some familiar workable perspective. Oswaldo's speech resumed its normal matter-of-fact pattern, and the men stood talking softly for a few minutes.

Without any warning, a thundering roar filled the giant dome and the air of the cave around them, a sound Turolla likened immediately to the Venezuelan Mono Grande roaring, now amplified to a nearly deafening pitch. They stood

34

unmoving while the sound reverberated around them, Oswaldo clutching at Turolla's arm, their cigarettes fallen to the ground. Less than fifty feet in front of them, across the beam of their lights, a tremendous boulder crashed to the ground, and the cave echoed with both the sound of screaming and the grating of rock against rock as more boulders fell. Turolla leapt to the left wall of the passage, dislodging some stones as he fell against it; Oswaldo remained standing in the middle, screaming back at the noise, firing his rifle aimlessly up into the blackness. He fired high up into the vault of the cave, as more and more boulders crashed to the earth in front of them, falling increasingly nearer to the spot where their two flashlights lay shining their beams along the right-hand wall.

Across the shaft of light, a moving form crossed, large and lumbering, preceding the rain of boulders and moving toward the two men. They fled, Oswaldo emptying his cartridge behind him as he ran, aiming without reason at the echoing sound of screaming and at the thundering rocks.

Reaching the lighted mouth of the cave, Turolla looked back to see if the huge animal had followed them into the light, but there was only blackness and the swirl of dust. Then again, as in Venezuela two years before, the roaring reached a crescendo and abruptly stopped. The thud of boulders against earth ceased. It was exactly 4:25 P.M.

In Turolla's clenched fist there was what he had thought in the darkness to be a flat cold stone; as Oswaldo mounted and rode away before him, he saw that it was a jadeite amulet, a stylized face in the shape of a small ax. He mounted his own horse quickly and rode after Oswaldo; at the Rio Chancis the two reined in together. Oswaldo's speech was again incoherent, and his hair was now gray in spots, with a large patch of it shock-white.

At seven o'clock the men reached El Picco Chiumpi,

where they'd spent the previous night. Unable to sleep or eat, they remained sitting upright, Pino telling Oswaldo stories of the United States and Canada, until he returned to his normal state. The change in his hair was something Turolla didn't mention; as the sun came up, the South American said he would return to the cave if Turolla wished it. Instead, they continued on.

Turolla believes now that the cave is not simply a cave of giant beasts, but, because of the amulet and the shaman's promise, he thinks it holds answers to the very questions of culture and prehistory that are his main interest and concern. It will take, he believes, a highly organized and well-equipped expedition to again enter the cave and find the answers he is convinced are there.

That the giant apes have something to do with these answers is something he also believes. What their relationship is, and has been through the years, is, until the cave is once again entered, a matter of infinite speculation.

As soon as the Moon and Sun had pierced the darkness, the Giants wished to observe the rising and setting of the Sun exactly. They went so far, til the sea held them back, there, where the white people live. At last they wanted to raise themselves up to the Sun and they determined to erect a tower. Whereupon the Lord of Creation said to the other Gods: "It is not right that mortals should lift themselves to us." So, with thunder and lightning, they destroyed the work of the Giants, who fled in terror. But the people who had until then spoken one language on the earth, were now all separated and began to speak in various tongues.

This translation of an ancient Mexican text, one of the earliest known, reflects that belief in the Age of Giants which is widespread throughout Central and South America. The parallels with the story of the Old Testament Tower of Babel are rather amazing, and the mention of the confusion of languages in this early pre-Christian context, as a result of the Giants' presumption, is one that appears in all the corners of world culture, suggesting an original tale that was much more than a simple allegory or fable.

Marcel Homet, in *Sons of the Sun,* relates another part of the Mexican cosmogony, concerning the four fabled prehistoric ages:

> In the first age, Tezcatlipoca turned himself into a sun and all the people of that time were Giants. But through their arrogance they precipitated the destruction of their age. Pitch black darkness ensued and the jaguars devoured the Giants. The Aztecs who were, historically speaking, much younger, looked upon the culture of Teotihuacán with its massive pyramidal structures in Northern Mexico as dating far back in the primitive history of a vanished age when Giants must have lived. For it was only these primitive Giants who could have erected such massive buildings as there are in Teotihuacán and the even greater Cholula pyramids. According to popular tales still current among the Mayas in Yucatán—so Meidiz reports—four Giants used to hold up the heavens at the four cardinal points. They are the four Bacabs and the story is that one of them is white in colour and his territory is the North, where the strong wind blows which carries everything eastward.

Homet then goes on to cite the work of Guayman Poma de Ayala, the picture-chronicler of the sixteenth century who also spoke of these four world ages. These pictograms illustrate the Giant myths of the Tiahuanacu civilization on Lake Titicaca, and concern the creation of the world. In

37

the first stage primitive man ruled; in the second, however, it was the race of Giants, who built stone houses for themselves.

"Also a coarse race did he [the World Creator] make out of painted stones" is what another version says and do we not recognize them in the gigantic statues in the ruins of Tiahuanacu? But, although they were covered with symbols and inscriptions they communicated nothing to us. Critical Father Cristóbal de Acuña, in 1639, in the sixty-third chapter of his report on the Amazon regions, wrote about the "River of Giants." This refers to the Rio Purus which comes down from the Andes, east of Cuzco. On its banks, among other tribes, there were Giants, the Curigueres, who, so the people who had seen them said, were sixteen hand-breadths tall.

The giants of South America, we can see, are older than an isolated tribe or a pair of screaming mutant apes. They are locked into a puzzle that resembles an enormous maze, with figures of gleaming gold around one corner and snarling foul-smelling beasts around another. The maze has many ways through which to enter, but, like all proper labyrinths, has only one exit. It may be that from one angle we see a gargoyle, and when we stand behind a different hedge, we see in the same place a god. Once upon a time there may have been no difference; once upon a time there may have been only one point of view.

In the capital city of Manaus, where, as we have seen, a whole squadron of fish-giants wriggles lustily along the Amazon, there is housed next to the ornate opera house an Indian museum exhibiting local folk arts and crafts. Alongside the pieces of earthenware and beadwork and shell crosses is perched a living exhibit, quite remarkable in itself. This exhibit is an ancient green parrot with age-mottled feathers and rheumy eyes who, upon prompting from certain of the nuns who maintain the museum, will sing, in a gravelly falsetto, a rendition of the "Ave Maria."

Far to the west, along the Andes' craggy cordilleras, his flight extending the length of the entire continent and the height of its tallest snowy peaks, is another bird entirely. This is the condor of the Andes, a fierce and powerful bird who, in Prentice's verse, "can soar, through heaven's unfathomed depths, or brave the fury of the northern hurricane." The condor, moving on thick and muscular wings, rhythmically opening and closing the slots of its gleaming feathers to catch the rising air currents, is a bird seldom seen by man. It is as yet not so near extinction as its California kin, who number only seventy, but it is rare. And it is mighty. It is a bird shaped for and from another time, with its ten-foot wing span patiently beating the Andean air, scanning with copper-coloured eyes the green and white earth below. It is ferociously beautiful and sings not one note.

The difference between these two birds and their respective situations is, of course, an obvious one. I would like to think that it is the silent soaring flight of the condor, rather than the brittle song of the parrot, that represents South America still. I would like to think of this flight as a metaphor for all the giants of this giant continent, but I do not know that the metaphor is any longer apt. The world grows smaller, increasingly man-sized, and what was once thought majestic and beautiful may ultimately be regarded as only grotesque.

The condor, unlike the parrot, is increasingly out of place, and it may be that, like the giant primate, he, too, is nearly out of time.

Three

One hundred fifty million years ago, a curious reptile crept out on the branch of a towering Jurassic tree. The wind blew lightly through the surrounding trees, but with enough strength so that a number of dead leaves and dried bits of grass whirled past the reptile's perch. The breeze was, in fact, fortuitous enough to prompt this smallish animal to relinquish his hold on the porous bark and, in a matter of exhilarating seconds, coast out into a whole new element and a whole new Class of beings.

He flew, and with this flight reinforced a physiological trend that would develop in various modes of acceleration through the Cretaceous, on into the Paleocene and Eocene epochs, spreading out in order and style so that by the Oligocene his heirs fluttered over the entire face of the earth, balmy as it then was. Even in the succeeding Miocene and Pliocene the bird population continued its proliferation undaunted; the warm breezes that had swelled the numbers up till then had not yet turned chilly in the Pleistocene and slapped icy shackles on evolving wrists.

But this first flier was something else. He was a reptile with aspirations, if you will. He was, most probably, a member of a small group of reptiles called the Pseudosuchia, whose inclinations were increasingly arboreal. Over the generations his body had mutated in one very important

way—the reptilian scales that covered him had modified into rudimentary feathers. The reasons for this modification and its success were directly linked to his life in the trees; in order to get from one to the other, he was able to glide, *using* the air, whereas others of his ilk who lost their claw-hold on the family conifer fell helplessly through it like so many scaly stones.

This fellow did not, however, evolve into that later group of flying reptiles, the pterosaurs, but instead moved along a more restrained but also more successful line of development toward becoming the first true bird of whom we have a fossil record, the *Archaeopteryx*. (The giant pterosaur pterodactyls later chose to particularize another style of modification, but never became true birds by making the change to warm-bloodedness and true-featheredness.) *Archaeopteryx*, while retaining its clawed reptilian fingers and tail with nineteen movable vertebrae, had as well all the requisite plumes to entitle it to bird-dom. Through its veins flowed liquid that had warmed itself to a high enough temperature to insure it the first place in what was to become the Class Aves nest.

Today's echoes of this classical hybrid, this early reptile-bird, come, like so many other echoes, from the Amazon forests. There, the young of the *hoatzin* species still crawl about through the branches and damp undergrowth, using those vestigial reptilian claws which have remained on the first two digits of their wings. According to Heuvelmans, this species of crested pheasant even when it has reached maturity doesn't cry out like other birds, but instead croaks like a frog and gives off as well a strong musky odour like that of a crocodile or of some species of turtle.

(The feathered serpent of Mexican mythology, called Quetzalcoatl, is perhaps not an altogether unlikely counterpart to this present-day hoatzin and to the earlier reptile-birds of the Jurassic era. As more and more discoveries are made, there appears to be less and less mutual exclusivity between what are thought of as imaginary and what are accepted as real and actual beings.)

Birds, whatever their adaptations, have persevered. The over-all evolution of birds has had as its major feature the efficient subordination of all other vital activities to that most important aspect of bird business—flight. In 100 million years the winged and feathered inhabitants of this planet have come up, in their various and specialized numbers, with amazingly efficient mechanisms for getting about. Even that handsome appendage, the penis, is just so much excess baggage to all male birds, except for ducks and ostriches and one or two other reactionary die-hard types. Those sex organs that are still around are put away on the physiological shelf until the season is right, at which time they become inflated like carnival balloons, are dealt with, deflated, and once again shelved until the next season. Even during the time of celebration, there is only one ovary and one oviduct in most bird females, and neither male nor female can be bothered with a bladder.

Obviously, a class predicated on such weightless efficiency has had little time for the lumbering, paddling, or swimming giants of other classes. Yet ever since the Eocene, there have been a few outsize moments. At that time, 90 million years after *Archaeopteryx,* a flightless bird nearly seven feet tall, called *Diatryma,* lurched up out of the swamps and abruptly vanished. At the same time, the elephant birds, of the order Aepyornithiformes, raised their own awesome, if flightless, wings. They managed to linger a great deal longer than the corpulent and ill-advised *Diatryma;* there are records of their existence in Madagascar as late as 1666, and some authorities consider that they were extant up and through the nineteenth century.

They were immense. Related to the ostrich and the emu, the largest species is thought to have weighed up to half a ton. From her skeletal remains, she appears to have been somewhere under nine feet in height—the reconstructed model in the Paris Museum is eight feet, nine inches tall. But she was heavy. Each of her eggs would have provided sustenance for a family of five Farafangana picnickers over

43

an entire weekend, holding as each did nearly two gallons of white and yolk. Whipped into a Madagascar mousse, her progeny undoubtedly met more than one such untimely end. She was the weightiest bird of all, ever, a thick-ankled giant whose existence and habits were first recounted to Herodotus and later recorded by Marco Polo. It is possible, in the light of various reports, that she may still stomp irately through the remoter swampy woodlands of Madagascar, like a silken-feathered hoarse-voiced Marjorie Main on the track of a splay-footed but no longer to be found Percy Kilbride. It is a sad thing to be unaware of one's own extinction.

The tallest bird ever noted, though not the heaviest, was the tasty giant moa of New Zealand, *Dinornis robustus*. She had her heyday in the Miocene, but managed to hang on through the Pliocene and Pleistocene, even leaving her as yet unfossilized remains in this century, tantalizing us, like her fat friend in Madagascar, with the possibility of her ongoing existence. *Dinornis* stood twelve feet high in her web-stockinged feet and enjoyed considerable fame and reputation in the mid-1800s. Moa bones still covered with flesh were found in 1823 on South Island, New Zealand, and tales of moa hunts and moa baiting there still abound.

But all these huge birds were basically warm-weather fowl. They managed to get by in an equatorial zone that, though chilly at times, never plunged them into the freezing bath of glaciation that numbed and then put on permanent ice so many of their winged cousins, nieces, and nephews.

As we move away from the girdle of the earth and travel toward her chattering teeth and her chilly toes, we see that these flightless giants could never have existed in zones where mobility was, and is, life. The arctic tern, a less sensational physical type than any lumbering moa but a remarkable bird nonetheless, travels, according to Lois and

44

Louis Darling, a distance every year of eleven thousand
miles from the Arctic Circle to Antarctic regions, and then
returns. The tyranny of verb over noun seems here, with
the tern, explicit; the wintry imperative is something like,
Spend your time in stretching your limits, not in refining or

restricting them. (If ever we were tempted to classify and categorize animals as different parts of speech, we would surely say that birds were verbs, and mammals nouns, with insects as adjectives and all reptiles and amphibians slow-moving adverbs. Fish, even considering their infinite variety, would still have to be merely punctuation marks, hanging somewhere between form and function.) The brain of such a bird as the arctic tern is worn on its wing, and it may in the long run turn out to have been a more satisfactory place than our own over-specificating cerebella chambers.

Poet Barbara Gibson's

Snow covers
Cold lovers

is a succinct appraisal of what happens to species or individuals whose energies and passions are tuned too low. Adaptation must, like passion, proceed at a hyperactive rate when coping with a plunging thermometer, or simply turn to ice. Strangely enough, large birds fare better than small birds in situations of extreme cold, due to the smaller ratio between their outside heat-losing surface and their over-all volume. Metabolic rates are necessarily brisk in the arctic zones, and to maintain what are the highest degrees of body heat in all the animal king and queendom—104 to 112 degrees Fahrenheit—takes some doing.

Cold-weather birds compensate. The carnivorous snowy owl, in addition to showing no profile at all against the white landscape he constantly cruises, has evolved outsize ears with eardrums so dissimilar in size and shape that their difference enables the owl to pinpoint exactly the vibrations of his squeaking night prey.

Adaptations and compensations aside, the birds of high altitude and heavy snows that do remain for any time on the ground may be involved in some inadvertent camouflage as well. Recently, one such bird has given rise to quite interesting speculation concerning primate giants and giant tracks. John Pollard, in a letter to the editor of England's *Country Life* in July 1970, tells of watching a pair of Alpine choughs, crowlike birds that range from Cornwall to the Himalayas, hopping determinedly through the Austrian snow. After each tandem hop, there appeared behind them a set of tracks which looked for all the world like a line of human steps. The choughs sank breast deep with each jump, and the tracks their movements produced were apparently remarkable. With a combination of downward plunge and outward flutter, the birds produced prints that were broad at the breast (ball of the foot) and suitably less heavy and rather fragmented at the tail (heel), but still feathery at the wing tips (toes).

The bird tracks (giant tracks) provide a whimsical note to the business of analyzing Yetis and Yeti prints; however charming Mr. Pollard's story is, though, it is hardly the definitive answer to the Himalayan riddle. But it sets the tone for a section of the world where the inexplicable is not necessarily problematical, and the smile of Buddha curls upward undeterred.

Tracks in the snow have engendered so much speculation and publication through the years that it is difficult to separate the validity of tracks from the validity, in the Himalayas at least, of the big-footed primates who may or may not have made them. I have already mentioned that ever since the turn of the century and Major Waddell's first mention of them, Yeti tracks have been dismissed as belonging to langurs, leopards, bears, human pilgrims, and wolves. Undoubtedly, a number of such dismissals is de-

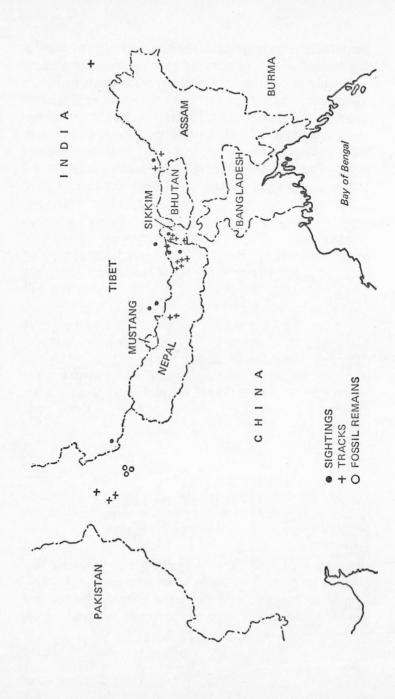

BURMA

ASSAM

Bay of Bengal

INDIA

SIKKIM

BHUTAN

BANGLADESH

TIBET

MUSTANG

NEPAL

CHINA

● SIGHTINGS
+ TRACKS
○ FOSSIL REMAINS

PAKISTAN

served. The Sherpas of the Himalayas have shown an obliging willingness to label all marks in the snow that come in pairs and are roughly man-sized as the tracks of the metoh-kangmi, or abominable snowman. And since the yellow-billed chough has been sighted at altitudes as high as its four-footed competitors in the Yeti-track sweepstakes, there is no reason to suppose that some of the credit for anthropoid impersonation should not be given to him.

But I must agree with Bernard Heuvelmans, W. Tschernezky, H. W. Tilman, and all the rest of the Yeti aficionados who believe that, in and among the wolf tracks and leopard spoor and bird hops, there exists as well five-toed bipedal proof of the real metoh-kangmi McCoy.

In the lately benighted country of Sikkim, high in the Himalayan clouds, there is a popular fantasy concerning the source of the sparkling clear water that flows beneath the coppery soil and gathers in bright green lakes among the mountain crags. It is said that the water comes from the mouths of great "snow lions" who hold up the earth and who live in the highest mountains of all. Only the holiest of the country's holy men are permitted to see them; they are white with blue manes, fire belches from their mouths, and they are troubled by no man, animal, or insect. The Nepalese, Lepchas, and Bhotias all recount this whimsical belief, with varying degrees of credulity. The mountains, clearly, are more than slabs of rock and ice to the people who live among them; they possess mystical characteristics as well. The highest peaks are still revered as the province of the gods; when a British expedition in 1955 scaled Kanchenjunga, at 28,208 feet, Sikkim's highest peak, the climbers did so only on the condition that they leave the summit itself unmolested. As good as their word, in fact, they refrained from scaling the last five feet.

It is not, however, at these highest peaks that the tantalizing footprints and sightings of the Himalayan Yeti have always occurred, but also in the slopes and valleys just below the summer snow line, where the melting ice nourishes the varicoloured primrose and iris, scarlet azalea, and thick dwarf rhododendron.

In the fifteen-thousand-foot-high kingdom of Mustang, some three hundred miles west of Sikkim as the chough flies, the air is charged with a panoply of spirits and malevolent demons. In the capital city of Lo Mantang, where only a handful of Westerners has ever been permitted entry, Michel Peissel, the French anthropologist, cataloged 416 demons of land, sky, fire, and water. Everywhere, once the sun goes down, appear lurking spirits which frighten the population, sometimes quite literally, to death. There are intricate demon traps set at the door of every house to ward these spirits off, and beneath each doorstep is secretly buried the head of a Mustang—the name comes from the words meaning the "plain of prayer," *sMon Thang*—horse.

Should any of these demons get past the horse heads and the ingenious wooden traps, should he somehow manage to appear and infect an unlucky victim, then that victim will quite certainly die, and afterward be disposed of in the usual Tibetan Buddhist fashion. Depending on which element he has previously chosen, or been born under the influence of, he will either be buried in the earth, cremated by fire, immersed in some icy river, or, should his element be air, neatly sliced into bite-sized bits and fed to vultures, in this way returning with them to the azure clouds of the Mustang sky.

Thorough and circumscribed are the religious traditions of the Himalayas. With this brand of Buddhism, the mountain day's activities invariably include the burning of religious butter lamps, the recitation of numerous prayers, the spinning of prayer wheels, and the unfurling of multihued prayer flags. Demons are everywhere, but so are depictions of the gods; in Lo Mantang stands a mammoth statue of Maitreya, the "Buddha who is next to come," towering three stories high over the low mud houses on either side of his temple.

Between and through these two heavily metaphysical states, stretching like the curved spine of a frigid but pliant serpent, the Himalayan Range exerts its silent and awesome, eminently physical presence. On the back of this beast, like fleas on a mastodon, the Yeti tales proliferate in the thicket of mystical belief (call it either religion or superstition—Himalayan belief has never been much interested in the often patronizing interpretations of the West). In Nepal it is common knowledge that to look upon the face of a Yeti is to die. There are tales of children and adults alike withering away after such an encounter. The Sherpas of Nepal and Tibet meet their Yetis most often while they are tending and guarding their yak herds at heights of up to seventeen thousand feet. Sightings by Europeans, since Howard-Bury's incident in 1921, have occurred at altitudes between twelve thousand and twenty thousand feet, a range where the atmosphere is unquestionably rarefied, and where both inner and outer vision seem remarkably keen.

There is no question but that some of these meetings have been later embellished by the addition of certain baubles and bangles of local lore. A number of stories probably never had a skeleton of truth beneath the legendary trappings in the first place; these are reminiscent of the accounts of that fanciful South American character, the curupira. In fact, there is even a similar detail concerning the feet of both myth figures, which John Napier cites in his exhaustive and impressive book *Bigfoot:*

51

J. R. P. Gent, another Briton, a forestry officer working in Sikkim, brought some extraordinary tracks to the attention of the scientific world. "The peculiar feature is that its tracks are about eighteen to twenty-four inches long, and the toes point in the opposite direction to that in which the animal is moving." The width of the track was six inches, Gent went on to say; "I take it that he walks on his knees and shins instead of on the sole of his feet. He is known as the Jungli-admi or Sogpa."

Three points about this seeming fantastic Yeti story are worthy of note. First, of course, Gent did not see the tracks himself but was merely reporting them second-hand from a native informant. Secondly, backward pointing feet is a widespread myth-motif, quoted all over the world in respect of giants and hairy wild-men. And, thirdly, the term Jungli-admi or "wild men of the woods" seems to remove this altogether from the realms of Yeti folk-tales. The Jungli-admi were undoubtedly humans. Colonel Stockley in *Stalking in the Himalayas and Northern India* (1936), refers to his casual coolie labour as Jungli. Major-General Macintyre in his book *Hindu-Koh* (1889), tells of a primitive, much-persecuted group inhabiting the forests and foot-hills of northern Kashmir and Nepal called the Jungli-admi, well-known to the British authorities.

(A fourth point might be that no animal except the African warthog goes about on its knees; this unsavoury animal does so only when eating, and has never been seen eating anywhere near either mountains or snow.)

So much for the Jungli-admi. They have no business on the floor while the Bigfoot Waltz is going on, any more than the fuzzy little curupiras of the Amazon. Also banned from the dance floor are the females of a Yetoid tribe who are reputed in the more fanciful annals of folk legend to possess anatomical features quite as extravagant as backward-pointing feet. It is said that these females carry

52

breasts so voluminous that it often becomes necessary for them to throw them back over their shoulders as they prance through the mountain azaleas.

Now, the real and actual Yeti has footprints which point in the same direction that he or she is bound, and boasts two perfectly respectable, but hardly excessive, dugs. (In lands where demons and gods are in constant competition, local conversation dwells as much upon the unseen as the visually verifiable. Therefore, just because a particular feature or characteristic is spoken of at length around the Sherpa campfire, its discussion is no proof that it really "exists." Exaggerated male and female features enjoy considerable interest and debate among all peoples, probably—I have heard the mythical proportions of John Dillinger's most lethal weapon discussed everywhere from Piazza Navona to Eaton Square.) In fact, the Yeti, even in its most demure appearances, is quite improbable enough without further gussying-up with whimsical or erotic detail.

We have already seen Bernard Heuvelmans' description of the Himalayan animal, formed from dozens of firsthand and, occasionally, secondhand reports concerning Yeti sightings covering the whole of the Asian mountain range. The evidence of footprints and sightings alike leads most observers to believe that there are two distinctly separate types of Yeti frequenting the different altitudes of the Himalayan slopes. The first of these, and the most well known, stands from five to eight feet in height and prefers the more inaccessible and higher range to that of the low snow-line range. The second is a smaller species, somewhere under five feet (rather more the size of De Loys's Venezuelan "monkey"), who is most frequently seen and reported in lower mountain areas where both the underbrush and the amount of human population are somewhat thicker.

The footprints of both the upper and the lower types indicate a pronounced large toe, not an opposable type of toe such as that found on other monkeys and apes, but a defined digit suited especially to plantigrade walking in the same way that our own human big toe is suited to our upright mode of getting about. The difference in the length of footprint between the two types of animal is as considerable as the difference in height, and more or less in proportion to it. The tracks found by the explorer and photographer N. A. Tombazi in 1925 measured between six and seven inches in length, whereas those discovered by S. Pranavanandra in July 1941 were twenty-one inches long. The mummified bodies of two Yetis that Tibetan lama Chemed Rigdzin Dorje Lopu maintains he examined in two separate lamaseries, one in Kham and one near Katmandu, were each approximately eight feet tall; Pasang Nyima described to Charles Stonor in 1953 a specimen the size of a small man, roughly five feet in height. This 1953 description corresponded to an earlier description Stonor had from a Pangboche villager named Mingma in 1949, which was made as the man watched the animal from his stone hut. Through a large crack in the wall he was able to see:

A squat thickset fellow, of the size and proportions of a small man, covered with reddish and black hair. The hair was not very long, and looked to be slanting upwards above the waist, and downwards below it; about the feet it was rather longer. The head was high and pointed, with a crest of hair on the top; the face was bare, except for some hair on the sides of the cheeks, brown in colour, "not so flat as a monkey, but flatter than a man," and with a squashed-in

54

nose. It had no tail. As Mingma watched it, the Yeti stood slightly stooping, its arms hanging down by its sides; he noticed particularly that the hands looked to be larger and stronger than a man's. It moved about in front of the hut with long strides. . . .

In addition to this over-all description, Mingma also expressed his awe at the size of the animal's teeth; apparently they were of a quite prodigious size. While he watched, the teeth were exposed to the man again and again as the creature repeatedly growled and bared them, much in the manner of a threatening baboon showing his long specialized fighting canines as an aggression display. There are other reports of the similarities between the Himalayan primate and those anthropoids of the more specialized ape family; perhaps it would be useful here to go over a few of them in comparison to the observed behaviour of other members of the hominoid group.

There are only four other known genus-limbs sprouting from the anthropoid tree (man, himself the fifth limb, excepted for the moment), and these are the orang-utan, the gibbon, the chimpanzee, and the heftiest branch of them all, the gorilla. Each of these, in varying degree, exhibits a certain likeness to the Yeti—and to the South American and North American varieties of large biped, as well—but no one among them really resembles him all that closely when they are made to stand up side by side for a brief inspection.

In all the print evidence gathered, there is no hint of the opposable large toe, the one pediform feature that links all the other anthropoids to their arboreal past and to each other. Every set of tracks in the snow, from altitudes as low as 9,000 feet to as high as 23,000 feet, reveals a down-to-earth digit that couldn't curl round a rhododendron branch in opposition to its other four fellows no matter how hard it tried. Even though some critics of the Yeti legend have dismissed its existence as merely a northern-reaching bit of orang-utan outcropping, there still remains the fact that no self-respecting orang would leave behind a trail with so little distance—or space of opposability—between the big toe and the second toe, and still call himself an orang-utan. And neither would any other ape. The orang, besides, is barely able to totter more than four or five paces upright before dropping to a simian stance and, once comfortable on all fours, making straight for the nearest Borneo bush, tree, or shrub.

The ape most able to negotiate long distances across land, and bipedally rather than on all fours, is the "graceful ape," the gibbon. Whereas the orang-utan is slow and cautious as he crawls through the Sabah scrub pines, and is downright clumsy out of their branches, the brachiating movements of the gibbon as it swings through the East Indian forest are as clean and efficiently beautiful as Stella stripes splashed against the leaves. On land, when fleeing across any open space to outrun a predator or an angry in-law, the gibbon lifts up its hands and arms in the manner of a praying mantis, swaying them gingerly back and forth for added balance as it runs. But the gibbon, though upright and graceful whether running or swooping through the trees, is never more than three feet in height and very seldom weighs more than twenty-five pounds. The gibbon is clearly not the stuff that Yeti sightings and footprints are made of. And the orang, even standing briefly and precariously upright, though larger than the gibbon, is not nearly large enough to stretch to even the low-line Yeti's stature. Furthermore, both the foot of the lethargic orang and that of the dainty gibbon could slip together with inches to spare into the smallest of Yeti glass slippers.

Now, the chimpanzee, although essentially a quadruped, can and does occasionally walk erect. And as far as his size qualifies him, he seems not at all an unsuitable candidate for the Yeti crown; he may reach a height of five feet, and even an inch or two past it, and may weigh up to 150 pounds. At that size and weight, he is not very far from the dimensions of some of the observed low-range Yetis. However, unhappily for those who would thrust this ape into the Tibetan spotlight, the chimpanzee has the same problem with toes and opposability as all the other apes; he spends three quarters of his life in the trees and is therefore primarily arboreal, impulsively bouncing about from limb to limb, and his feet show it. Though he may be able to walk for some distance mainly on two feet, his manner of movement actually necessitates leaving four tracks in the earth rather than two; like the gorilla, he supports his weight on the toughened skin pads below the knuckles on each hand, and not solely on the bottoms and sides of his feet. He moves most comfortably bent slightly forward, his arms and hands stretching out before him like props, his fingers curling deftly inward. (Still and yet, there are the fascinating experiments of one Professor Sydney W. Britton. Placing a chimpanzee that had never before been exposed to or even seen snow on a field where several inches had recently fallen, Britton was surprised to see the animal, after its first few tentative steps, stand up and walk erect in the snow. This evidence might very well suggest walking habits quite different for the same species depending on the different conditions of their climates. The assumption is that, in the Himalayas, Andes, Rockies, or Cascades, an ape could have gradually learned, or been conditioned toward, walking upright through the snow in order to reduce the painful and numbing contact, on two of his limbs at least.)

Nevertheless, no matter what the conditions surrounding this change to an upright stance, the tracks would still reveal a walk conducted on widely spaced toes, like the other apes', quite as opposable as they could possibly be.

If a chimpanzee can raise itself erect, abandoning its usual stance and gait when placed in a foreign troublesome en-

1 *Frame from Patterson film of 1967 showing female Sasquatch near Bluff Creek, California*

Roger Patterson

2 Collections of casts of Sasquatch footprints belonging to John Green, October 1973. The cast he is holding is a duplicate of a cast made by Roger Patterson in 1967 in northern California; it measures 17 inches long by 7 inches wide

3 Giant footprint (13 by 18 inches) photographed by Eric Shipton on the Menlung Glacier, Himalayas, in 1951

4 Tracks in the snow near Mt. Adams, Washington, 1969

Robert Morgan

5 Print from Skamania Country Road, Skamania, Washington, made during National Wildlife Expedition, 1970

Allan Facemire

vironment, then perhaps its African compatriot, the mountain gorilla, is capable of doing the same thing in a similar situation. The gorilla, who walks in much the same fashion as the chimp, leaning forward to support his great weight—often over four hundred pounds—as he moves timorously through the Rwanda and Uganda highlands of Africa, might be induced to throw back his massive shoulders and stride manlike through the wild celery stalks, were it not for one interesting but restrictive anatomical feature. This feature, in fact, is what separates a man's walk from an ape's, and has to do with where he sits. Man has a strongly developed buttock muscle, the gluteus maximus, which as he walks pulls his body forward and over his leg with each step. The gorilla has none of our cushy tissues, however, and is unable to compensate for any such stride forward with a pull backward. Furthermore, the bones of man have managed to take the pressure off the leg muscles so that when he is standing erect they lock together at the knee. The gorilla is unable to so lock his legs, and the pressure of standing upright is borne painfully by the quadricep muscles on the front of his thighs. Walking in an upright position is even more uncomfortable than standing in one; therefor his shuffle is understandably oblique.

Nevertheless, the reaction of Professor Britton's plucky chimpanzee in the snow is an intriguing one and raises more questions than it answers—the main question being why should a primate stand erect for the first time. The question is not an altogether easy one to answer, inasmuch as there are several interrelated factors involved. It is generally thought that when apes first descended from the trees, the prime reason for their change in habitat was that they could not ignore the temptation of the food supply spread out before them on the ground. The berries, roots, and insects that littered the flanges of the then-tropical forests of Europe, Asia, and Africa made up a smorgasbord that became, apparently, too inviting to pass up. Gradually, in tentative leaps and curious bounds, down they came. As they lingered for longer and longer periods between trees and below trees and even, occasionally, completely out of sight of trees, they were far from being as relaxed as they were when in those trees. In this new and unfamiliar situation, this ground-level feeding, their eyes were kept thoroughly and constantly peeled for predators; the newly grounded apes found that by keeping their eyes as high as possible their optic defenses were optimal. It is theorized that as they craned their necks and strained their limbs to get a peek over the low grasses and savannah shrubs, they got used to the idea of tiptoed vigilance and never went back to their previous vulnerable crouch.

But such erectness had more to do with ground life than simple vigilance in the face of unfamiliar dangers. There was also the question of the use to which arms ought to be put once they'd swung a body cautiously down to the grassy flatland from the leafy bough. Already, ape arms and wrists and fingers were rather more dexterous than their monkey-neighbor counter-parts; the centuries spent in brachiation had specialized their reach to a quite admirable degree. It was an extension of this already established dexterity that put these same fingers and hands to the task of poking into the earth for edible roots, up-ending and scraping moist rocks and stones to get at the juicy beetles and fat grubs that clung to them, and snatching up any unlucky toad, frog, or other smallish creature that ran before them through the grass.

This specialized use of ape fingers on the ground was one thing, but the next step toward ongoing bipedalism was quite another; the first *Ramapithecus* who picked up a bit of twig and idly pushed it into the Pliocene mud to squish a beetle or cut a root, with these motions also shut behind him the creaking door of four-footedness forever. As we have seen in the twentieth century, once a tool or weapon is picked up, it becomes nearly impossible to ever put it down again. So at that time, the ape with a tool became the ape with a purpose. There came about very quickly a process of learning and specificating that is called by scientists today the process of positive feedback. What happened, we assume, is that the ape, once aware of what his fingers could do as they grasped a bit of twig or thumped a bit of granite, stood even taller as his hands changed functions before him. For a creature suddenly absorbed in tools and their possible uses, the development of the brain became increasingly important.

(This business of tool-handling and heightened awareness is, one must point out, a matter of purest speculation on the part of present anthropologists looking for an answer to the burgeoning-brain riddle. There is logic in it, of course, and a certain pattern of parallel processes between tools and brains. But it is still speculation.)

Cause and effect are here somewhat mucked about; the impetus toward enlarging and refining one's thinking processes theoretically comes from the act of standing up and using in a new way a part of one's anatomy, but at the same time the impetus toward standing erect in the first place and using hands and fingers in a new way comes from and through those very brain processes. If there were thrown into the hopper of "mutual feedback" yet another motive/process factor—the painful contact of hand to snowy ground, say, or the existence of certain berries only at shoulder level or above—this back-and-forth process would be even more accelerated, and the learning ape would tend to stand, think, and make use of his fingers at an ever faster rate than before.

At this present moment in time, there are no known apes north of either Sumatra, in the case of the gibbon and the orang-utan, or the Sahara, in the case of the gorilla and the chimpanzee. But there are ape fossils throughout the entire Europe-Asia land mass, from Siberia to Helsinki, in areas that have gone through climatic changes infinitely more varied than the changes currently occurring in the present ape homelands. All this bit of theorizing means to suggest is that perhaps warm-weather apes who found themselves sometime late in the Miocene up to their fuzzy ankles in drifting snow may have pulled their hands out of the drifts and set about their various brain-enlarging tasks with greater alacrity than they would have otherwise employed. And, as we have mentioned with birds, the larger the body surface and the greater the bulk in cold-weather climes, the better; high metabolisms in tiny creatures have a way of burning themselves out all too quickly.

Thirty-five million years ago, in the early Oligocene, the earth's weather began to boil and freeze and then boil again with such nasty unpredictability that the vertebrates of the time were forced to go through a number of changes that otherwise they would have undoubtedly eschewed. These weather shifts were the crucible out of which our present life forms were tempered and re-formed, and the living population of the present world is a very different dish because of this early session of pressure-cooking. But the less dramatic changes of the late Miocene were in their own way as crucial to the formation of certain genus traits as the earlier upheavals.

Imagine a furry primate, swung down from the withering branches of a dying tree in the midst of what he regards as a summer storm. Imagine that primate scratching about in a numbing white substance, extracting with difficulty the corpses of small insectivores and the dead branches of chilled blueberries. As he eats and moves across the wind-swept savannah—be it in present-day Omsk, Siena, or Dijon—he keeps his two front limbs pulled up close to his body in order that his hands and fingers will remain warm and usable. Imagine him finally dying, and imagine as well most of his offspring dying in the cold—until there are very few of him on the land mass that is Europe-Asia.

But imagine survivors. The great-great-grandsons of the original almost-ape, who have trotted gingerly across the hardening and then melting ice, have managed to squeak through. By now, those who remain have developed strong enough muscles in their thighs and hips to support their weight without their bodies having to stoop. By now, their fur has thickened, and the layer of fat on their bodies has given them a heftiness their ancestors would have found threatening, if not obscene.

The sun never really melts the snow, but by now it doesn't matter. In their veins and chests, changes have occurred; millions of red corpuscles have been added to capture the scarce oxygen molecules in the thin air, and there are concessions to the high-altitude solar radiation as well. The only problem is, there seem to be so very few of them around. But they go on, trekking across the now-familiar snow, in certain seasons venturing down into what seem to them the uncomfortably hot lower slopes and humid valleys, eating whatever there is to eat, maintaining somehow their vast bulk.

There are barely enough of them to mate, and not nearly enough to form any sort of community. It is each one for himself, and singly, or in occasional pairs, they continue their egregious rounds through the snow. With their hands they methodically scoop out sections of moist earth and slowly digest whatever life they find growing in its mineral loam; they continue, and never once remember nor ever once revert to, the practice of going about on all fours. They have not done so since they first learned to walk through the dead forests, not since the snow and ice became facts of life, not for at least thirteen million years.

Conjecture regarding how and where and why the Yeti

came about, and what the long years before and since *Gigantopithecus* have meant to him, can be only honest conjecture at this point. But conjecture and a few teeth here, and a femur or two there, are the beginnings of paleontological theory. As the discoverer of *Gigantopithecus* himself, G. H. Ralph von Koenigswald, once remarked, the fragments of pre-Neanderthal man available for inspection and study could all be comfortably assembled on one medium-sized table. Since then, in 1956, there have been more teeth and bones unearthed, but fossil evidence is never plentiful for any hominid, anywhere, at any place, in any time.

The living Himalayan giant is a frustratingly seldom seen creature, but out of those sightings that have taken place there emerges the barest outline of what his habits are. And behind him in the snow he has left, as well, not only tracks but little piles of clues to his diet and eating habits. Charles Stonor, pursuing one such line of tracks, twice came across droppings that contained fur, rodents' bones, and earth. Another investigator, Gerald Russel, in analyzing his own collection of droppings, came up with the following fecal breakdown: "a quantity of mouse-hare fur; a quantity of mouse-hare bones (20); one feather, probably from a partridge chick. Some sections of grass, or other vegetable matter, one thorn, one large insect claw, three mouse-hare whiskers." The reports of bearers and mountain natives have it that the animal is fond of marmots, mouse-hares, and large insects, and also that he consumes substantial quantities of earth, perhaps for the mineral value or perhaps simply for bulk. There are also a few reports of young yaks, tahr, and musk deer crossing his palate.

Depending on the time of year, the food supply over twelve thousand feet would vary considerably, and the Yeti eating regime would vary with it; in late fall and

65

winter he could make do by subsisting largely on the fleshy likes of hare and occasional yak, while in summer he could enlarge his menu to include a variety of greens, plus roots and berries, and the kernels of available mountain grains.

Professor René von Nebesky-Wojkowitz writes that the natives of Tibet and Sikkim include yet another element in the Yeti diet—salt. There is a particularly saline moss which grows on the rocks of the moraine fields, and it is for this moss that the animal spends a considerable amount of his time searching. Finding it, and satisfying his salt hunger, the Yeti is able to return to his more secluded region of habitation once again.

Von Nebesky-Wojkowitz also includes in his account the observations that Tibetans, Sherpas, and Lepchas have made to him concerning other than dietary matters:

> He fears the light of the fire and in spite of his great strength is regarded by the less superstitious inhabitants of the Himalayas as a harmless creature that would attack a man only if wounded. . . . From what native hunters say the term "snowman" is a misnomer, since firstly it is not human and secondly it does not live in the zone of snow. Its habitat is rather the impenetrable thickets of the highest tracts of Himalayan forest. During the day it sleeps in its lair, which it does not leave until nightfall. Then its approach may be recognized by the cracking of branches and its peculiar whistling call.

Though the Yeti was not originally thought to be a nocturnal species, the evidence now seems to point to that possibility. A respectable case might be made for the infrequency of sightings during the day because of the animal being generally sound asleep while the sun shines.

Whether a night creature or a day creature, the Yeti seems by most accounts to harbour none of the ill-will that the Monos Grande of Venezuela and Ecuador expressed against Count Pino Turolla. The nastiest reports have always to do with a good deal of roaring and leaping about, but there are no reports in the Asian files of anyone's ever being done in by one. (Those deaths that have occurred have been the result of natives allegedly looking into the face of the animal, not of having met with actual physical harm at his hands.)

In fact, the Yeti is by nearly all accounts a solitary beast. Like the orang-utan with whom he has been frequently confused, he carries with him an air of almost melancholy apartness and of morose solitude. Without the physical and psychological bolstering that a pack of his fellows backing him up would give, he engages in little of the serious marauding that his primate relatives in other regions do. Unlike the baboon or macaque, both of whom have developed highly aggressive habits since they first dropped from the boughs, and whose social patterns reflect the cohesion of that ongoing aggression, the Yeti is never seen in packs, and his behaviour is therefore never the sort that can be called group-oriented. There is only one observation on record where more than one animal was seen in the same place at the same time; in *The Long Walk,* by Slavomir Rawicz, there is a detailed description of a pair of the animals encountered by the author and his group of escaping prisoners, on their flight from a Siberian prisoner of war camp across the Himalayan slopes. The physical description of the two animals, one slightly larger than the other, pacing back and forth in a circular shuffle before the group of escapees, adds very little to the descriptions already noted; the point of the incident is that a pair of animals was observed, rather than the usual solitary beast. Whether, as the author surmises, this was actually a male-female pair, or two males, or two females, is impossible to know. There are bachelor bonds, and all-male

67

groups among gorillas, baboons, and chimpanzees; this pair may have been such a one, with a larger and a smaller individual, out for a midday stroll.

The difference in size between the high-land and low-land types of sighted Yetis may conceivably be the difference in size of males and females of the same species. The varying shades of hair colour observed could be a reflection of the variety of age or the change in seasonal adaption within one species. Comparing the sexual dimorphism of those other two land-master primates, the macaque and the baboon, once again, we see that the males of both these genera are twice as large as the females and coloured quite differently as well.

It is therefore not inconceivable that the large and the small Yeti, called the *Dzu-teh* in the case of the first, and the *Meh-teh* (or *mi-teh* or *yeh-teh*) in the second, are in reality only the eight-foot male and the five-foot female of the same species, or perhaps the eight-foot-tall adult and the five-foot-tall juvenile at different stages of growth. Possibly, there is only one brand of Himalayan primate, after all.

Himalayan legend has it, however, that there are indeed two separate species and that there have always been two, since time out of mind. There are even those who say there was once a third variety as well, one that ranged from thirteen to sixteen feet in height and possessed jaws so large and powerful they were able to crush the head of a yak between them like a walnut. This giant's giant is said to have wandered about the mountains in groups, high above the snow line, capturing and consuming all available prey that presented itself at altitudes over thirteen thousand feet. A Tibetan lama named Punyabayra gives the name of this immense creature as *Nyalmo* and maintains that the Tibetan people have always known of his existence. The lama's description of the other types, the *Rimi* (the Dzu-teh of Nepal) and the *Rackshi Bompo* (the Meh-teh), corresponds nicely to the Nepalese descriptions—even to the point of their affection for the Barun Khola Valley there, where they are said to be both well-tempered and numerous, if elusive still.

Back across the mountains—in the land of demons, burial by the element of one's choice, and the sound of whirring prayer wheels—the tiny kingdom of Mustang harbours, a link with the giant past that, thus far in paleon-tological history, has yielded up no present answers. On the faces of the towering cliffs that form the barren wind-swept Mustang mountains, there have been lately discov-ered hundreds of chiseled caves, great whistling mouths in the stone that inside them contain networks of tunnels and rooms with dozens of hidden passages connecting them one to the other. The entrances to these mountain caves strongly evoke the caves of Ecuador and Peru, but are more numerous and seem to have been formed with a sense of community about them. Michel Peissel has counted twenty-nine of these cave clusters, each of them with over fifty separate stone mouths apiece, and each well up out of the reach of modern man. The entrances are tool-carved and tool-shaped, rather than naturally formed, and are placed so far above the present ground level of the sur-rounding plain that exploration of them can be carried out only with the aid of intricate cliff-scaling equipment.

The native Lo-bas have no explanation for the caves or their history; in a land teeming with both natural and supernatural explanations for every possible phenomenon, this absence of legend is somewhat strange. The caves are large enough to suggest the use of not unconsiderable strength and force in their creation—a good deal more muscular power than is found in the arms of present-day Mustangers, at least—and also considerable antiquity. Perhaps the tribes of giant Nyalmos, who, according to Punyabayra, once roamed the mountain slopes, crossed over from nearby Tibet and used these high huge caves as their shelters from the snow and ice. Or perhaps they are the works of an early true-man in an early unrecorded epoch. Whatever past they do conceal, and whoever were their long-ago inhabitants, these caves will one day be opened up with archaeological precision, and the prehistory of the Himalayas will be richer for the effort. Even as they now stand, mute and tantalizing high above the treeless Mustang plain, they are as so many stone asterisks on the testament of what is known about both the past and the present in this harsh country. The clues are everywhere so large, and the available answers so very small.

Before we move again across the oceans and into the third great giant range this book concerns, there is a story I have heard a number of times on the psychic grapevine that I would like to here relate. It has to do with the two mountain ranges we have so far explored, the Andean and the Himalayan ranges, and of their relationship to each other and to the rest of the world and its history. It concerns first things, and the religious significance of them both.

It is said that early on in the history of the world and of religions, when, in fact, a great deal more of the earth was covered by water than the present five sevenths of its surface, there were set up at different times in the two

70

highest mountain ranges two centers of religion and culture in an otherwise cultureless era. The reasons behind choosing each of these sites had to do with their respective positions as earth chakras—chakras being centers of physical and psychic energy—on the earth at that time. That time, of course, was long before the physical persons of either Buddha or Jesus Christ had appeared on the earth; it was a time when the animal-human line in most areas of the world had not yet been drawn, a time when the chalicothere and the megathere roamed, and when mankind was still an idea as yet unthought in a pre-australopithecine brain.

The chakra site of this very first center was not in the then-swampy Olduvai Gorge nor in the caves of the Dordogne or Lascaux; it was located instead in the highest peaks of the high range of the Andes. At this site, on a continent whose edges are very different now from what they were then, communication with and from the "gods" first began.

The Andean center was chosen by these "gods," or extraterrestrial colonizers, because it afforded them the easiest "window" on the world. They chose much as today our own NASA flight engineers choose particular sky sites when and where the spatial window is open and accessible. The Andean site was the site of maximum energy contact in a time when the oceans and the plains refracted and distorted the energy focus.

Chakras, in the teachings of present-day Buddhism, denote wheels of karma, and so represent the never-ending round, called samsara, of human striving and human failing which is created as the spirit passes from one incarnation on to the next. Just as there are metaphysical wheels, there are real and physical wheels of energy as well; these energy centers occur on the body of the earth, and also on each individual human body. There are glandular counterparts to some of these human points, and the practice of acupuncture may one day be revealed to deal with others. But these chakra centers change as the body changes, either due to natural progress and processes—in the case of the earth as it passes through the different energy fields of the planets and its axis shifts—or due to the more violent upheavals that occur upon and beneath the face of the earth itself. As a man or woman matures, the chakras of his or her body make certain subtle adjustments; a violent accident, however, will shuffle and rearrange the energy points with the same rude jolts as an earthquake will shuffle and rearrange mother earth's.

Such a shuffle and shift accounted for the change in the position of this planet's heaviest chakra site in the Andes; the energy spotlight, either gradually or abruptly, consequently swung around and across the seas until it focused on what we now call the Himalayan Range.

When this change of focus and energy took place, a number of other things occurred as well. Those evolving humans who'd been terrestrial guinea pigs in the garden center now were left alone there to function without their instructors high in the Bolivian and Ecuadorian clouds. They continued for some time in the expectation and hope that the gods would return, going on with the tasks they had been assigned. It was not that they were simply abandoned by their teachers; the process took enough time so that both aliens and men were prepared for the leave-taking when it happened. The men were set free to function on their own, responsible now to themselves as much as to the code of behaviour they'd learned, using that code and their increased awareness. The actual physical shift of chakras may have been linked to the kind of earth upheavals that have destroyed subsequent Andean populations, and if such a process of physical upheaval took place, there may have been very few of these animals-with-thoughts, these proto-men, who survived.

In the Himalayas, meanwhile, those primates who were thrust into the chakra light reacted as well as had their South American counterparts to thought and to speech; whatever means the alien gods used to increase their subjects' cerebration and communication processes had their effect, and Asian consciousness was dramatically raised. There was set up, as time went by, a repository of knowledge, a record of the first beginnings at both sites, which endured for tens of thousands of years. In the jungles of Africa and the fields of Iowa during this same long stretch of time, less-enlightened and less-favored man made do as best he could with obsidian arrowhead and bamboo pole.

High in the mountains, the most advanced of the aliens was referred to by a name passed down from instructor to instructor, the "priest from across the sea," or the dalai lama. It was the task of the newly specialized men who inhabited the chakra center to maintain the record of knowledge concerning the original center and the original colonization. It was necessary to guard the center and that record from whatever forays might be made into the remote settlement by other evolving primates, who were then as savages to the enlightened mountain priests. After a time, the alien gods disappeared from the second center, as around it on the earth's surface the gulf between civilization and savagery gradually lessened.

When Gautama Buddha, Prince Siddhartha, appeared on the earth, he appeared as the descendant or representative of the ancient alien gods, as a direct manifestation of that god spirit and instruction. Ancient knowledge became collected in the vessel of Tibetan Buddhism, and the vocabulary of Buddha became the vocabulary of the initiated. Five hundred years after Buddha, Jesus Christ appeared, himself an avatar of the original teachers, and, through the Essenes, became again as one with the ancient knowledge.

Before and during the long periods that preceded these men, the chakras of the earth shifted subtly and then shifted again. There were eras when diffuse chakras shone in Lemuria and Atlantis, and there was finally the return of the heaviest focus of all to where it had first resided, the South American Andes.

Whether this second great and final shift had occurred much before the discovery, in the 1660s, of the existing dalai lama in Tibet is not known. Neither is it clearly known what those practices and methods of enlightenment were which had been carried out in the two centers over the thousands of years since the initial colonization, nor why the colonization occurred on this planet in the first place. All the explanations, plus the records of the civilizing mutations that were induced in the minds of the first men, are said to exist now in only one repository of knowledge, in the place where the true dalai lama is also said to reside. It is guarded, as it has always been guarded wherever it lay by the devout and the enlightened, not far from where the deeds it records took place centuries of centuries ago. It is kept in a community where speech is irrelevant and where every action approaches the sublime—in a remote mountain monastery in the Andean cordillera, in that unassuming temple that is called the House of the Seven Rays.

Four

In the ongoing history of life on this planet—the kind of limited history we deduce from fossils, rather than from myths—there has always been a fluctuation of vertebrate habit and habitat with regard to the elements. This swinging pendulum of preference started when the fishes first crept out of the sea and their gills began to do new and unheard-of things with oxygen and carbon dioxide, and continued through that time we have earlier mentioned when the first reptile-birds edged out on their Jurassic limb. Though elemental preferences continue to change (watching that change through the eons is rather like taking the boat-train from Paris to London: forging through the Channel seas, soaring high up over those seas at Dover, then plowing on through the good English soil, all in the same body, tucked in between the plush ribs of the same sooty beast), the first element that nourished and nurtured life was the same one that tends the majority of life's representatives today—water. At the time, admittedly there was a great deal more of it around, and the options were limited—sustenance was taken where it was found, and it was found in the primeval ooze or it was not found at all.

But earth and air gradually became elements to reckon with—Heraclitus and salamanders are the only ones who ever believed in fire as an alternative—and it got to be more and more possible for developing life forms to experiment a bit. There evolved ever more tangential ways to be an animal—or a plant, for that matter—as more and more differentiated combinations of environment and gene occurred. It was, in those early days, a system of radiation and bounce and of nearly surrealistic combinations; tooth went sometimes with claw, sometimes with hoof, and feathers and scales were seldom more than a few generations apart. That which flew also swam, and occasionally ran at a gallop through the towering forests as well.

Animals were not the only life forms stretching their limits; plants did some highly original adapting along with them, and these plant changes made possible even more elaborate cape-work by the animal hordes. In fact, had there been less innovation in the plant world in those early eons, we might never have gotten around to our first primate steps. Flora and Fauna joined together at the hip of mutual stimulation and dependence in a way that enabled both to perform and grow. The veneration of the religions of the world for tree totems is a deserved one; from Midgard to Eden, whether Trees of Knowledge or of Life, or the bodhi tree which sheltered Buddha, the relationship of man with tree, animal with plant, is one of total interdependence. It is not merely the interdependence of oxygen and carbon dioxide, but a more far-reaching balance of life systems that puts us forever in debt to the greenery around us. It is the nature of the plant itself that concerns our own generation and makes life as we now know it possible.

Before there were flowers there were trees. The early trees were scaled to the dimensions of life at that time, life ruled by the clammy paw of the giant brontosaur and two-footed tyrannosaur, as well as by the rubbery wing of the huge pterosaur. The tree world, when the first preprimate was still rehearsing his prosimian lines and mulling over in his minuscule brain whether or not to enter the primate arena, was immense to all but the giant reptiles who lumbered through it. To that first pre-shrew tree shrew who is today understood to be the progenitor of our line, the world was as large and unworkable in size as our modern steel and aluminum forests must seem to any hapless city mouse in their midst. The pavements of the forests then were no more carpeted than our own sidewalks—it was a time before grass, a time when the giant conifers dropped their brittle cones to the brittle earth with a clang and a thud. Wind had to blow and rain to fall, or else the conifer spores would wither and die where they lay, unstructured as they were for any type of adaption.

It was with the rise of the angiosperms in the shadow of these great pines and spruces that plant life at last began to assert itself and a new and ingenious relationship grew up between leafed and feathered creatures, between flower and fur. Where the pinecone before had held dry naked spores as it fell heavily to earth, the new angiosperm created and held as it grew an altogether new invention, a seed encased in a flower. Compared to the rigid conifer spore, the seed thus created was amazingly self-sufficient. All it needed was a little help in getting about—help it enlisted from mammal, bird, and insect in all their proliferating variety—and if that help wasn't immediately forthcoming, this new seed could bide its time till help chanced to come along. It had its own nutrient shell which sustained it, a shell that became an excellent energy source to those animals who chose to eat it rather than transport it.

Forty million years or so before the first mousey-primate, the first angiosperm rustlings began to be heard on the plains and in the forests; by the time that primate arrived on the Paleo-scene, Ms. Nature had switched over the majority of her plant systems to this new scheme of flowers and seeds, and all, or nearly all, her eggs had been firmly deposited in the angiosperm bank.

Life proceeded at a gallop. Over and across the swamps and fields and forests, as new flowers grew up, new animal species grew up beside them to munch on their stems and leaves and in the process receive a few clinging seeds on their fur and whiskers. These seeds traveled to new sites, where new species sprang up and spread them. Insects and birds—for now there were real birds flitting through the trees where before the giant flying lizards swooped—also carried the seeds and flowers, and the air as well as the earth filled with life. Pinecones were turning to pomegranates and peach blossoms, and the leaves of the new trees were certainly tastier than any had ever been before— the change in the epiglottal structures of those animals who bridged the two eras must have been considerable; pine needles can never have been an easy mouthful. Hard palates softened and bird teeth changed to bird beaks, and between and among the new species, the pre-prosimian maneuvered himself upward on the mellowing bough.

It turned out, for a number of reasons, that this small mouselike creature took to the trees with the same facility with which his later descendants, the apes and ape-men, left them. He was comfortable there, out of the way of the clumsy toxodons, nibbling on a succulent leaf while they nibbled on each other. Stalks, shrubs, grasses, bushes, vines, flowering and fruited trees—everywhere the greenery beckoned, sending out a wispy tendril here, a pale blossom there. And everywhere the mammals of the earth and the birds and insects of the air responded; the process of plant-eating produced a meat-eating correlative—herbivores bred carnivores with sabre teeth, and the energy of plants and seeds passed from the flesh of one to the flesh of the other.

Meanwhile, the cunning prosimian, in the face of all this scurrying activity, remained in the trees and got on with the business of filling his stomach and fitting his furry little body to the branches, producing a handsome diversity of progressively more prosimianiacal progeny as the centuries slipped by beneath him.

Alas, he was probably not, after all, a very likable little fellow.

These millions of years later (roughly 60 million), there is a living representative of the primate order who differs only slightly from this first mousey climber. The tree shrews of Southeast Asia still dangle somewhere between an insectivore past and a primate present, while looking, we imagine, almost exactly like their Paleocene forebears. It is a curious thing, but their behaviour resembles as much that of the ultimate primate, man, as it does the behaviour of any other quadruped or biped in between. These frenetic little tree shrews are much given to excess in all their appetites—gluttons and libertines and psychopathic wranglers we might call them if we were given to anthropomorphic license—and their seemingly unwarranted belligerence goes well beyond the demands of either territoriality or purely sexual behaviour (whatever purely sexual behaviour can possibly be). They are nasty little buggers (once thought to be exclusively insectivorous), and if they were the size of man, they might well give him aggressive pause. As prototypes of the primate alpha, they are distressingly close in their competitive habits to the primate omega—it is as if, somehow, the genetic-behavioural pattern were sketched out in harsh chiaroscuro as a preliminary cartoon for the canvas of mankind, now ameliorated with all our gradations of texture and hue, but nonetheless as true to the first charcoal sketch as to the last elaborate swirl of paint-daubed brush.

But it is the period in between shrew and man that is the concern of this chapter, a long stretch of time that has to do with getting used to life in the trees and then getting unused to it. In between the branches, there appears to have dropped down a heavy and rather more well disposed primate than we are generally wont to meet. This variety of primate is the one who managed to sequester himself and his heirs on remote mountains and in dense forests while the business of aggression was being carried on by his genetic siblings. That he once existed there is no scientific doubt. Twice as large as the largest gorilla, he roamed over a range whose geographical and temporal limits are as uncertain as the terrain that ultimately shaped both his stature and his habits. It was a varied landscape over which he shambled. In order to place him as accurately as we can on a chart of physical development, we might do best to go back out onto the limbs of the trees in the Paleocene forest and follow the progress of the primate line, so far as it has been theoretically reconstructed, and so far as we are able to say we know it. If we like, we can think of the primal tree mouse, the first little scrapper, as the inhabitant of the lowest of all available branches. With this image in mind, let us, like him, endeavor to climb upward and outward, scampering into and out of each new tree-house identity as it presents itself to us. The Paleocene is where we begin.

During the great proliferation of life, which took place in the trees as much as on the ground in those old halcyon days of the early Eocene, the prosimians bred with an alarming alacrity and considerable variety. After their initial venture upward, the first-class adaptation of life among the branches was an immediate refinement of that means to increasing specialization, the grasping mechanism. The first shrewlet, as he moved upward, relied on his claws and

not his fingers to pull himself along; his heirs, without
notable exception, relied instead on what they themselves
developed—grasping opposable digits which they were able
to wrap around branches, squeeze grubs with, and carry

out a hundred other tree tasks with, in a number of then-innovative ways. They invented and used verbs unknown to the ground-focused insectivores who had preceded them. At first their newly acquired digits all moved together, as they do today in the motions of the primitive marmoset's hand as it gropes about the South American bush. Once clung, a finger couldn't be unclung until and unless the entire hand agreed. This principle is rather like the one governing birds at rest; as a starling settles in for the night on the bough, her claws lock together and around the branch as her legs relax into a slightly stooped sleeping position. Any subsequent nightmare of cats or hawks may send her spinning backwards and under her perch, but will never dislodge her from it, so long as she neither wakens nor stiffens. She may flip over like a duck in a shooting gallery, but fall she won't. The early prosimian grasp was as thorough as the starling's; later on it relaxed a bit and delegated decisions to each discrete finger.

Prosimians filled the limbs with furry life, moving higher and higher on grasping digits with each subsequent generation. Over sixty now-extinct genera left their fingers pressed in fossil remains, and the surviving representatives of the prosimian line continue their insulated lives in Madagascar, some twenty additional species strong. (It is only on this isolated island, though, that present-day prosimians still flourish, removed from their heirs and from the majority of twentieth-century stress. They are called lemuroids, perhaps after their dreamlike, ghostlike manner, perhaps after the legendary continent of Lemuria, above whose sunken shores they now range.) As the more progressive prosimians grasped their way upward, they also looked increasingly alert. Their noses shrank and their eyes enlarged over the centuries as they replaced ground sense with tree sense and

6 Skamania Country print cast, held by Robert Morgan

Eve Phillips

7 Jim Butler with LEFT *North American print, and* RIGHT *Eric Shipton's Himalayan print*

8 Jim Butler's foot in The Dalles, Oregon, sheriff's office cast

Robert Morgan

Robert Morgan

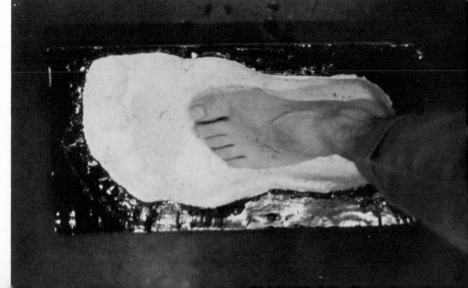

9 Pueblo rock paintings Abo Settlement, Abo, New Mexico

10 Jadeite amulet (5 by 7 inches) found by Pino Turollo in 1970 expedition to Guacamayo Range during Mono Grande attack

Robert Morgan

Pino Turollo

11 A giant ape killed in 1924 in Venezuela by de Loys

François de Loys

adapted their sensory apparatus to the needs of their new environment.

Some among them remained the size of insectivores, while others grew to proportions very grand indeed. The *Megaladapis* was the size of a small ox or cow, with immense jawbones slung under his face, flattened cheeks and brows, and extremely long and, we suppose, agile prehensile fingers. Getting about from limb to limb took cautious doing for the *Megaladapis,* but he managed to survive and thrive, along with his slightly less bulky genera-companions, for a goodly span of time before he disappeared.

The twelve-inch-long tarsier, meanwhile—a still living single species that boasted two dozen different genera 50 million years ago—evolved in a different direction from that of most monkeys, equipping himself with attenuated leg bones (tarsals) so large and springy that he was able to hop with ease a distance of over six feet and land on a dime, or the Eocene equivalent thereof, with considerable precision. He also developed the largest eyes and ears of the entire order and took to hopping about at night, rather than during daylight hours, thus becoming the first nocturnal primate.

The indri was also developing at the same time as the tarsier, but he chose more monkeylike ways, while never actually abandoning his prosimian status. Today the indri darts about on his hind legs, resembling an upright monkey with the face of a fox or dog—he is the only prosimian, and indeed one of the few primates, who goes about on two feet. The female, unlike her sisters in the Madagascar ghetto, carries her young clinging to her belly and not her back, and behaves generally like a prosimian with simian affectations.

In the balmy West Indies there remains a group of early prosimians who have persevered into this, the late-late Pleistocene. They are the lorises, and they include species so diffuse as the bush baby—who out-tarsiers even the tarsier in length of bounding leap—and the family of lorises, which seems to have been founded on the principle of cautious sloth. These lorises have a basal metabolism so low that, like a Miami Beach matron, each would wither away and die if deprived of her fur coat.

Then there were, and are, the pottos of West Africa, who decided to give up their index fingers in order to widen their grip, and to refine one of their toes (called the grooming claw) for the purpose of grooming their immaculately gleaming, if wiry, coats of hair. The aye-aye, too, gave itself over to a rather kinky bit of specialization; the middle finger of its small hand has left mere grasping opposability behind and turned itself into a kind of pencil-sharp probing wire—the better to dig out grubs and beetles with, from the shards of bark that its chiseled and protruding rodentlike teeth have exposed.

These various prosimian embellishments, however imaginative, are ones that the next group of upward-scrambling primates as a group decided to bypass; the primates intent on monkeydom eschewed wiry fingers and excessive gambadoes and opted for more tasteful refinements somewhere around 31,000,000 B.C.

Monkey-specificating took place as much within the body as without—sight centers in the brain enlarged markedly, and as they enlarged they began to differentiate colours and refine the incoming image through stereoscopic means. This meant seeing as no previous primate had ever seen, in depth and with colour variety. The other senses were altered as well. The snout ends, now reduced in proportionate size, gave up their highly sensitive tactile hairs, and their hands instead took over the functions of touch and feeling as they became more and more specialized in both function and structure.

Later, certain of the monkeys moved on into apedom when they began using these highly developed hands to even more particularized advantage and developed the process of brachiation. Brachiation, or arm-swinging, was a means of getting around and into feeding areas that otherwise would have been too impossibly fragile to explore. The monkey method of feeding was to come down with a graceful, and sometimes not so graceful, plunk, dead on top of whatever limb they were interested in—this was the approved method of prosimian feeding that all the primates up till then had engaged in. What the soon-to-be apes did was to swing in at their food instead of bouncing down on top of it and pluck it from the limb from beneath or from the side. After this seemingly simple change of habit, the new breed of tree-dweller was able to handle papaya and mango that had been previously beyond his grasp, munching all manner of nuts and fleshy fruits from otherwise inaccessible vines and boughs.

Apes, unfortunately for paleontology, have left precious few fossil records. It is theorized that they first began doing for arms what monkeys had previously done for hands, at a time somewhere between 25 and 30 million years ago. This body-building regime bulked them up considerably— the primate aesthetic went through quite a few changes as it dropped the idea of a lean and hungry look and replaced it with a more robust and well padded image. It was again, as with big birds and big fish, a question of a large powerful body requiring a less frenetic metabolism to keep it going. As the ape muscles thickened, the newly muscular primates could afford to relax a bit more than their monkey predecessors. Perhaps that is why they were not so intent on populating the mangrove and banyan groves to such a zealous degree as their immediate ancestors were. Whatever the reason, the apes, then and now, are few compared to the number of monkeys and men; they may have reached too high a level of food-gathering efficiency too fast to throw out the alternative lines that would have led them into other, perhaps less successful, branches of development.

The time between monkey and man is in any event a somewhat more gentle time than we sometimes realize. The world's prehistory is not all volcanic fire and brimstone; for every reign of climatic terror, there is a corresponding period when Marie Antoinette sits languidly fanning herself at Versailles, enjoying her pastoral pleasures as the clouds around her dissolve and thicken. During the Miocene, that period of time between 25 and 13 million years ago when the earth was filled with the pleasant sound of grazing mammals, and when the great ice sheets were not yet even fields of snow, there occurred a primate development that is little recorded because it is as yet little known. At that time a number of apelike creatures fell to earth like the overripe pieces of fruit on which they fed, softly, without creating any great calamity in the branches they had left, nor in the grassy fields over which they presently began to roam.

These were big-bodied folk, considerably larger than the *Ramapithecines* who at the tail end of the Miocene were also dropping in groups from the boughs. These large pithecine ancients, bulked up by brachiation and encouraged downward and outward by an unknown quotient of curiosity, are the ones who, in my lay opinion, immediately preceded *Gigantopithecus*. Like enormous cookie crumbs leading across the savannah, these heavily biceped and triceped apes formed a path of development that ended at the doorstep of a house in whose windows are still seen the flickering shapes of the Yeti, the Sasquatch, and the Mono Grande.

Whereas the old primate shrewoid was better off fend-

86

ing for himself in the trees—out of the way of the new angiosperm-induced grazers, foragers, and predators on the ground—this latest beefed-up ape was not about to be stepped on and squashed underfoot by any careless megathere or mastodon. He was able to take his pleasure with ease, and his ease with an increasing amount of pleasure —at least in the beginning; later on complications arose. But at first, long before the smaller *Ramapithecines* (who are considered to be the oldest genera to which the label of humanhood sticks) began their own process of switching over from leaf-munching to grain-chewing habits, these larger than arboreal life types were sampling their first helpings of cereal—stalks of oat and wheat and barley— and liking the taste well enough to stay on the ground for course after course.

Elwyn Simons and Peter Ettel have speculated that the line that produced *Gigantopithecus* must have passed directly through the lair of *Dryopithecus indicus,* a largish ape from the Miocene and Pliocene, who may have also been ancestral to our present African gorillas. This may be so, but I rather think the proportions of the great-granddaddy *Gigantopithecus* were heftier than those of *Dryopithecus,* massive as he was. Although the line that led up to this ancient giant is admittedly as difficult to trace as the subsequent line that led down from him to our present-day beasts, we do know that such a line had to exist—that the emerging giants and giantesses hardly appeared, full-blown and full-fledged, out of nowhere.

The earliest appearance on fossil record, earlier than Von Koenigswald's original teeth or the subsequent Chinese discoveries in Kwangsi, comes with the jawbone of a barely pubescent female who met her maker in what are now the Siwalik Hills of India, and whose age is estimated at somewhere between 5 and 8 million years. Before her, there is only surmise. She had her antecedents, of course, and they provided her with extremely muscular jaws and graminivorous teeth, as well as a frame designed to keep predators at bay. Strangely, considering her size, she was no predator herself and never developed a taste for flesh and blood. Those brachiating apes who grew large and then grew curious, and who came down from the trees to scratch about in the briar and grain patch, never took to the ways of their later smaller cousins who found in meat and meat-eating a whole new reason for living—the manufacture and use of tools. They, as the first progenitors of the giant race, simply grazed, taking their cue from the ruminating pattern of the day. Until their own bones and back molars turn up, though, there can be very little theorizing as to how large they really were or as to how they came to be so thoroughly and comfortably at home on their newly acquired range. We had best concern ourselves with their great-great-granddaughter instead—the teen-ager from the suburbs of New Delhi who is the oldest giant around.

In the late Miocene and early Pliocene (from 12 million to 6 million years ago), the foothills of the Himalayas, where this teen-ager romped, were given over to just the sort of vegetation and climate that her jaws indicate: dry grasslands filled with energy-giving seeds, rhizomes, and cereal grains whose constant chewing gave her a muscular jaw and a mouth crammed full of broad flat-crowned teeth. As with the current apes, and man, these teeth numbered thirty-two (the same number as De Loys's anachronistic monkey): peglike incisors, large canines, and thickly enameled molars, all crowded together to a much greater density than that which they achieved in the mouths of the previous tree-rovers and papaya-suckers whose method of mastication followed an entirely different pattern, with less grinding and more tearing and shredding.

The Indian jaw, found by one Sunkar Ram and turned over to a Yale-Panjab expedition in 1958, and the Chinese jaws found ten years later in Kwangsi province, combine to indicate proportions twice as large as any present-day mountain gorilla. These proportions are much more similar to those of the Mono Grande of South America, the metoh-kangmi of the Himalayas, and the Sasquatch of the Pacific Northwest than they are to any other primate, real or imagined, before or since.

When fully adult, the *Gigantopithecus* male weighed somewhere around six hundred pounds and stood over nine feet tall. The female was shorter and correspondingly lighter, in keeping with the scale difference in land primates today. Judging from the evidence of the jawbones so far discovered and the more than one thousand isolated teeth that have been unearthed, decanted, dislodged, and otherwise revealed, both male and female walked upright and chewed mightily.

The Siwalik Hills teeth are prodigious. It would seem, from the lack of a cutting edge on the two incisors, that there was no carnivorous shredding or tearing of meat in the *Gigantopithecoid* repertoire; the menu kept its yin-yang balance intact without recourse to either flesh or blood. A macrobiotic diet reinforced the prevailing macrodontic structure; though the outsize teeth ground exceedingly slow, they ground up sizable enough amounts of plant protein to keep the nine-foot-tall metabolism humming. *Gigantopithecus* thrived on the prairies, and her range extended certainly the width of present-day China and India, pushing out on all grain-swept borders, and possibly over a much broader area. The bones left behind in New Delhi and in the Hong Kong outback attest to a far-ranging existence; the climate then was different enough from the climate now that any present limits to giant wanderings would at that point probably not have applied. As their smaller but fiercer primate cousins multiplied around them, it is probable that a good bit of range-extending took place in reaction. An interesting question is which end of the Chinese-India range was crossed first; as the animals moved to the meter of competition and food supply, and extended their range into and out of Europe, were they heading toward, or just possibly fleeing from, their first range, the range of North and South America?

The question of which continent-home came first—North America, South America, or Asia—is not one that enjoys much of an audience with giant buffs and anthropologists, generally; Asia has pounds and pounds of evidence to its credit, and, so far, the Americas between them haven't enough to put in a nickel bag.

Grover Krantz, the anthropologist at Washington State whose work on Sasquatch lore has done much to make the present North American creature respectable, has reminded me that the Bering land bridge, between Alaska and what is now the Soviet Union, is the only access route since the Eocene that connected Asia and America. He speculates that *Gigantopithecus* could have been a relatively scarce, though geographically widespread species in Asia, whose extension into North America would have been automatic if he were able to survive as far north as about sixty-six degrees latitude. This was not an especially treacherous survival adaption, as we can see from those mammals who made it a means to their migration, and who have since moved back and forth with considerable ease—bear, bison, wolves, and man.

Since there have as yet been no teeth or jawbones found in either North or South America that would indicate a habitation as old as that in the Siwalik Hills of India or the plains of the Kwangsi province in China—since, in fact, there have been no *Gigantopithecus* fossils or teeth at all turned up here—there can as yet be no supportable claim made for the idea of an ancient giant roaming the Andes or Cascades before he ever roamed the Himalayas. The only claim that may be made is that it is possible that the ancient giant, like the camel, began his wanderings in the Americas and gradually worked his way northward and eastward, finally arriving in China and points farther south, instead of moving across the Bering bridge in the opposite direction. That it is not totally impossible is the only claim I make; other animals bucked the tide of what is now considered the general pattern of migration, and it may be that as we find out more about what the continents were really like in those days, we will find out that not all movement followed what we now deduce to be the majority line. Man included.

But probability will certainly have to wait for a fresh supply of jawbones and molars and canines and will need more than one or two isolated fibia to stand on. The case for a giant anthropoid primate living in North America some 4 or 5 million years ago is a matter of sheer conjecture and empirical perversity until archaeological substantiation in one form or another turns up, whether in the La Brea tar pits or the shales of the Mississippi delta. But the case for a giant anthropoid primate stalking about, alive and well, in this century is a different matter altogether, a case where the eye-witness reports of its appearance, habits, and diet

91

are legion, and where its footprints, as fresh as the ground they're impressed in, are rife.

Sasquatch, to put it mildly, is a monster aficionado's dream. Like a bulky will-o'-the-wisp, he/she has tantalized the good people of the Pacific Northwest for over a century, and at present shows not the least sign of slowing down or fading back into the scrub pine and mountain ivy.

The first stories, like the first stories of so many things, that are told about Sasquatch are told by the Indians. The name itself is a Salish name—used by those Indian tribes of southwestern British Columbia who refer to a creature in their forest pantheon as simply "the wild man of the woods." In the Klamath Mountains of Northern California, Sasquatch is called *Oh-mah-'ah* and, more popularly, *O-mah* by the Huppa tribe and its descendants there. Northward in the Cascades, where mountain peaks stretch up to heights of fourteen thousand feet along the seven-hundred-mile length of the range, the name of the apelike giant is *Seeahtik,* and his home is said to be in the deep thicket of Vancouver Island.

Other names, from other tribes of the Pacific Northwest and California, are *toki-mussi* and *gilyuk* and, rarely, *hoquiam.* There is a tableau in a dark corner of the American Museum of Natural History in New York that portrays three Nootka Indians in what are described as ceremonial costumes and masks; behind the glass is a figure representing "wild man," completely covered with hair and with a face very like an ape's. The brow and chin line are sloped like an outsize orang's, and the nose is short and flat. An interesting feature of Kwakiutl wood carvings and masks, depicting this same creature, is the mouth—it is pursed in a whistling position with pouted lips, ostensibly making the high-pitched and powerful screams its real-life counterpart emits as it prowls through the conifer forest.

John Green reports that stories of giant ape-creatures abound in the small islands off the coast of British Columbia, where vegetation is extraordinarily thick and high, and where primitive conditions still prevail. At Klemtu and Bella Bella (over fifty-two degrees north latitude) there are reports of apes which have been seen swimming from island to island, or going about their business on the sandy shore, either singly or in occasional family groups. These, and other Indian reports of sightings and of footprints, go back a century at least and are an integral part of the local legend. It is said that in times past, which means times before the white settlers came, the Sasquatches, like the Nyalmos of the Himalayas, were far more numerous and ranged over far wider areas than they do today. (John Green, in his trailblazing *On the Track of Sasquatch,* estimates from his own research that the range of tracks which have turned up in the period of the last seventy-five years extends over an area more than a thousand miles in length and half a million square miles in area.)

Early references to the northwestern giant come from explorers and trappers in the territory, and they are early indeed. The first written report, aside from Indian reports and legends, concerns tracks and comes from David Thompson's *Narrative,* a condensation from the diary of one surveyor and trader who explored the headwaters of the Columbia River at a time when Alexander I sat on the Russian throne, George III rested on England's, and, in between the two, Napoleon Bonaparte ruled from a position that owed a great deal to some rather extensive exploring and trapping of his own. The United States of America was at the time a strapping thirty-five years old.

Here is that journal's entry:

93

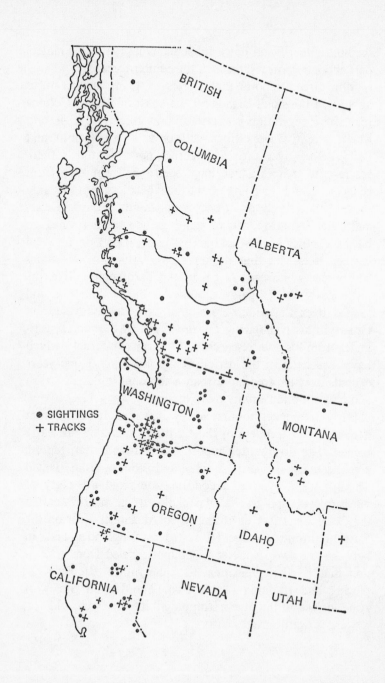

SIGHTINGS
TRACKS

January 7th. Continuing our journey in the afternoon we came on the track of a large animal, the snow about six inches deep on the ice; I measured it; four large toes each of four inches in length to each a short claw; the ball of the foot sunk three inches lower than the toes, the hinder part of the foot did not mark well, the length fourteen inches, by eight inches in breadth, walking from north to south, and having passed about six hours. We were in no humour to follow him: the Men and Indians would have it to be a young mammoth and I held it to be the track of a large old grizled [*sic*] Bear; yet the shortness of the nails, the ball of the foot, and it's great size was not that of a Bear, otherwise that of a very large old Bear, his claws worn away; this the Indian would not allow. Saw several tracks of Moose Deer. 9PM (Ther-4)

This report, from 1811, was written by a man who later was the first to be credited with traveling the entire length of the Columbia River and who subsequently mapped most of western Canada and the northwestern United States. He was no fool, certainly, and none too credulous; his matter-of-fact description does both the tracks and him justice.

The oldest report of an actual sighting of a Sasquatch, as opposed to the discovery of tracks, occurred in the Coast Range, south of San Francisco, sixty years after Thompson discovered his footprints in the snow. It appeared on November 5, 1870, in the weekly Butte *Record,* and contained the testimony of a man from Stanislaus County who, while camping in the mountain woods, met up with the immense creature. At the end of each daily hunting excursion, the man would return to his campsite to find the remains of his fire strewn about, ashes and burned log ends scattered on the ground. Nothing else at the site was ever disturbed except for the fire bed; some three hundred yards from the camp he also came across a series of human tracks in the wet sand, "bare, and of an immense size." The man resolved to find out the cause of his fire's disarray and the identity of the track-maker, and set himself up in the brush seventy feet from the fire and waited through the day for a number of hours for someone to appear. Someone finally did, preceded by the sound of a shrill whistling call.

The creature, whatever it was, stood fully five feet high, and disproportionately broad and square at the fore shoulders, with arms of great length. The legs were very short and the body long. The head was small compared to the rest of the creature, and appeared to be set upon his shoulders without a neck. The whole was covered with dark brown and cinammon colored hair, quite long on some parts, that on the head standing in a shock, and growing close down to the eyes, like a Digger Indian's.

As the man watched, the animal approached the campfire, looking back and forth toward the forest and the surrounding brush. He threw back his head and whistled once again and then reached down and grabbed a stick from the still active fire. He swung the stick around his head a number of times, until the fire went out, and then reached in and did the same with another smaller branch. He continued for a good fifteen minutes, methodically extinguishing the blaze and scattering the ashes about. After completing this operation, he walked back out of the camp for a short distance and returned with a companion, a female, and the pair, always walking upright, passed within twenty yards of where their observer crouched. It seemed to him as he watched the two of them that their aim was somehow the pleasure they gained by their frenzied monkeying about with the fire; he reports they took a kind of delight in swinging the pieces of wood around and around through the air.

This first recorded encounter set the tone for later reports and sightings; the animal seldom showed any ill-will, just a good bit of curiosity and respectable primate behaviour. The early reports are few, however, and reflect the difference in attitude between what an observer may without fear of ridicule relate he has seen, and what he may not. This first California report, in fact, came after another vague report of a "gorilla" sighting was referred to in the press, and opened on a rather defensive tone, expecting the reaction to be none too credible. The earlier reports, from the middle of the nineteenth century, still reflected the territorial optimism and egotism of the time, however, and were related with a kind of feisty believe-it-or-not bravado. During the late 1800s, though, and the early twentieth century, this openness closed up like a Victorian clam, and the man who spoke of seeing anything freakish out there in the timber was the man who automatically exposed himself to a considerable amount of comment on the Great Northern platform. The last report that appeared in the climate of Victorian self-consciousness was one penned by no less a forthright and unconcerned chronicler of the West than Theodore Roosevelt himself.

In his book *Wilderness Hunter*, Roosevelt tells the story, in 1892, of a man named Bauman, and his experience fifty years earlier. The man and his partner had been trapping in the mountains between the Salmon and Wisdom rivers in Idaho and found themselves in an area that was considered dangerous because of a killing there the year before. Another trapper had been slain, and his body half eaten by some type of unknown predator, and the area was now an infamous one. When Bauman and his partner

97

returned to their campsite the first day, they were surprised and vexed to find their camp in a state of considerable havoc. Their packs and provisions were torn apart and scattered over the ground, and the earth was covered with tracks, tracks the two men did their best to explain away to each other as those of a bear or some other four-footed beast. That night they were awakened from their sleep by a large animal crashing into their lean-to shelter, smelling mightily of what Bauman called "wild beast odour." Bauman shot at the looming animal but missed his mark in the darkness, and the beast retreated through the low brush. The next morning huge biped tracks again covered the ground; the next afternoon, after a day of laying traps and hunting, the men returned to again find their camp torn apart, their shanty destroyed, and their kit and bedding again tossed about. The next night, the men stayed awake until dawn, keeping watch before the fire, rifles firmly in hand. They were audience to the animal's "harsh, grating, long-drawn moan," a peculiarly sinister sound that came from the hillside opposite their camp. The sound of the animal circling the site kept them alert throughout the night, and the next morning they decided they had had enough of their marauder and would pack their provisions, gather their traps from the previous day, and leave the place. Bauman went about the task of pulling up three beaver traps in a nearby lake; the job took him longer than he'd intended, and he sent his partner back to camp to make the final preparations for their departure while he finished with the traps and animals. After two and a half hours of dealing with the trapped beavers he returned to the camp. It had grown late, later than they'd planned, and Bauman was worried about having enough daylight left in which to travel; nearing the edge of the glade where the camp lay, he shouted as he approached, but got no answer back from his friend.

The camp-fire had gone out, though the blue smoke was still curling upwards.

Near the fire lay the packs, wrapped and arranged. At first Bauman could see nobody; nor did he receive an answer to his call. Stepping forward he again shouted, and as he did so, his eye fell on the body of his friend, stretched beside the trunk of a great fallen spruce. Rushing toward it the horrified trapper found that the body was still warm, but that the neck was broken, while there were four great fang marks in the throat.

Printed deep in the soil, the animal's footprints told what had happened. Bauman's partner, after he'd finished up the packing, had sat down facing the fire, to warm himself and wait for his friend's return. He had apparently been seized from behind, his body broken and thrown about.

The animal had not eaten the body, but apparently had romped and gambolled around it in uncouth, ferocious glee, occasionally rolling over and over it; and had then fled back into the soundless depths of the woods.

Bauman, too, fled, abandoning everything but his rifle, making his way out of the glade and mounting his tethered pony by nightfall, and then riding straight through the night. The terror stayed with him all his life, along with his resolve to never again enter that Idaho forest.

It's an ugly little story, this one, and the only one I've come across that ever gives Bigfoot such a thoroughly malicious character and such a bloodthirsty mien. He sounds here much more akin to the South American Monos Grande than to his precursors or heirs in North America or, for that matter, in the Himalayas. He is atypical, and this atypicality is fortunate—his character, in this incident

Roosevelt reports, may have been that of an unhinged and solitary individual, already ostracized from his own pack or tribe, a renegade even among his own kind. Or it is possible that he was not so unique after all and that his behaviour represented the style of the day among his fellows. If this is the case, then it is apparent that these days the style has noticeably mellowed.

Through the years, the Sasquatch reputation has softened, for there has been no evidence discovered, nor reports made, concerning the kind of homicidal behaviour of this early incident in Idaho. Sasquatch descriptions may vary, and the range of observation fluctuate, but the main body of evidence, like the body of the beast itself, remains overwhelmingly benign. The worst that can be said is that occasionally both bodies do give off an unpleasant odour.

Aside from one such questionably scented report made by a group of trainmen in 1892 who captured what they called a smallish ape (four feet, nine inches in height), in Yale, British Columbia, there are no other recorded glimpses of either apes or ape-giants until well into the twentieth century. (Jacko, as this black-haired, long-limbed fellow was dubbed by his captors, immediately disappeared from view after his first mention in the press, and what he was, if he was, is now impossible to guess. Strong and reticent, but given to a sporadic sullen bark, as he was described he sounds rather like a chimp.)

In 1924 a variation of the man-bites-dog story occurred when, instead of being himself the captor of one, a man found himself being carried off into the bush by a full-grown animal. The man was Albert Ostman, and his story as presented by John Green is remarkable in its detail, while intriguing in its ramifications.

The Ostman story begins in the manner of many early Bigfoot scenarios, with a hunter-camper alone in his isolated camp. The camp was in an area of British Columbia famous for lost gold mines, the area at the head of the Toba Inlet. Ostman had been brought to his site by an Indian guide, who told him about a prospector who was said to have been killed in the area, and also spoke of the presence of large hair-covered people in the mountains there, wild men who left tracks (such as the guide's uncle had seen) two feet in length.

After a day of scouting around, Ostman set out for higher ground, and the Indian left him; he reached an altitude of one thousand feet and spent some time prospecting in the damp rocky earth. He found no minerals of interest and the next day moved on over rough country to a spot with a small spring, where he settled in for two more days of digging and hunting. Leaving this site, he continued over more and more isolated terrain until he finally made permanent camp near a spring that faced a grove of young trees by a high rock wall. The first night in this grove passed uneventfully, except for his discovery the next morning that some night visitor (which he assumed to be a porcupine) had uncovered his provisions and, while not taking anything, scattered them all over the ground. He spent that day hunting and fishing and returned to camp at dusk, made his supper, and settled down for the night. The next morning he found a half-pound package of prunes and a box of pancake flour missing, and his packsack overturned and emptied out on the ground. He stuck close to camp that day, watching for intruders of any sort, but saw none; that evening he went to bed with his rifle tucked beside him in his sleeping bag, determined to stay awake all night in order to see who had been disrupting his camp.

He fell asleep, however, and it was that night, after a week and a half in the forest, that he was kidnapped.

He awoke to the feeling of being lifted off the ground and

101

squashed into his sleeping bag; the top end was drawn and held tightly closed over his head. He was then carried over a long stretch of uneven ground—so it seemed from the bumping, jolting pace—and then across smoother, then again uneven terrain, for a distance of three hours. Unable to either use his gun or draw his knife because of his cramped position in the bag, he wasn't able to loosen his cords until he was finally dumped on the ground by his captor and unceremoniously set free.

His legs had become so numbed from the journey that he was unable to stand up; he was only able to assess his situation from a half-seated position, and what he was able to make out in the darkness was more than a little surprising to him. His captor now joined a group of three other figures that he was gradually able to distinguish in the early morning light as they stood softly chattering nearby. He realized that they were the sort of creature his guide had spoken of, the hair-covered giants of the mountains. He describes, in his sworn account, a family of four—an adult male, an adult female, and two younger ones, a boy and a girl. The boy, whom he estimates at between eleven and eighteen years of age, was approximately seven feet tall and weighed about three hundred pounds:

> His chest would be 50–55 inches, his waist about 36–38 inches. He had wide jaws, narrow forehead that slanted upward round at the back about four or five inches higher than the forehead. The hair on their heads was about six inches long. The hair on the rest of their bodies was short and thick in places.

The older male, who had carried him over the mountains, was considerably larger than the "boy," with teeth longer than the others' and eyeteeth longer still, but not long enough to be called tusks.

The old man must have been near eight feet tall. Big barrel chest and big hump on his back—powerful shoulders, his biceps on upper arm were enormous and tapered down to his elbows. His forearms were longer than common people have, but well-proportioned. His hands were wide, the palm was long and broad, and hollow like a scoop. His fingers were short in proportion to the rest of his hand. His fingernails were like chisels. The only place they had no hair was inside their hands and the soles of their feet and upper part of the nose and eye-lids. I never did see their ears, they were covered with hair hanging over them. . . . If the old man were to wear a collar, it would have to be at least thirty inches.

The proportions of the two females corresponded to the differences between the two males. The older of the two

was over seven feet tall. She would be about 500–600 pounds. . . . She had very wide hips, and a goose-like walk. She was not built for beauty or speed.

The females had longer hair on their heads than did the males, and this was curled slightly upward, as if in a kind of fringe of bangs. The younger female, unlike the older one, was virtually flat-chested.

Once the morning light had become bright enough for Ostman to habituate himself to the movements and appearance of his hosts, or abductors, he was able to see that they lived in a kind of small valley of eight or ten acres, bounded by high mountains. At the southern edge was a V-shaped opening, eight feet wide at the bottom and twenty feet at the widest point at the top; he assumed it was through this opening that he had entered the valley, and through it he would have to make his escape. On the east side wall of the valley, in the side of the mountain was carved out a cliff or shelf with a sizable outcropping of rock above it. Ten feet deep and thirty feet wide, it contained what Ostman later was able to determine were blankets woven of narrow strips of cedar bark and packed with what appeared to be clumps of dry moss. These primitive blankets were used by the group the first night and during subsequent naps, while Ostman slowly consumed the provisions he had left in his pack (which the large male had brought along with the sleeping bag and its contents).

The family of four were not brilliant in their divertissements. Ostman was kept in their company for nearly six days, barred from leaving, but at the same time unable to figure out just what it was they wanted from his company in the first place. They seemed, over the hours he spent with them, to be extremely attentive to his behaviour, but not overly anxious to engage him in any kind of meaningful dialogue. In fact, what they were in the long run most interested in was, not his mind nor his body, but, rather, his can of snuff.

It was this tin of snuff that provided a means to Ostman's escape from his family of Sasquatches. From the first day of his capture, Ostman noticed that the large male showed a marked interest in what and how he ate. When he took out a pinch of snuff and placed it in his mouth, but neither chewed nor swallowed it, the "old man" seemed fascinated by this departure from normal eating procedure. At last, over Ostman's morning meal of coffee and hardtack with butter, the decisive pinch was taken, and the animal ceased battling his own curiosity—he simply reached out to the prospector, grabbed the nearly full box from his hands, and emptied the entire supply into his mouth. After he'd

104

licked clean the container as well, the sticky resinous substance began to do its work on the animal's innards.

Rolling about, he set up a squeal "like a stuck pig" and became quite violently ill. He was rendered so powerless, and his family in their anxiety was pushed to such distraction, that Ostman was able to pass through the hole in the rock wall which had until that point been kept guarded from his approach. He fired one shot into the air as he fled —the only time during the whole encounter that he ever used his gun—to frighten back the larger female who, gathering her wits in the face of her mate's indisposition, suddenly ran after him as he fled, shrieking in agitation. The shot frightened her back inside the wall, and Ostman was able to continue his flight, running as far and as fast as he could, until at the end of three hours' running he came in sight of Mount Baker. After pushing south until he was too exhausted to continue farther, he spent the night in the open, and the next day stumbled into a logging camp on the Salmon River branch of Sechelt Inlet. From there he was able to make his way back to Vancouver.

Ostman's description of his adventure has by now become a Bigfoot classic; there is even a recent R. Crumb comic-book strip, "Whiteman Meets Bigfoot," which in gloriously funky detail retells the tale in more updated, and considerably raunchier, terms than Ostman himself ever reported.

But the original Ostman narrative, even though it contains each and every itemized provision that he took with him, and mentions what he shot and what he ate, up to and after his abduction, still provides only a limited description of the giant animals' actual behaviour. The females were generally meek, he says, and the adults did a lot of resting. The young male was given to playfully bouncing back and forth on his rump, with his docile sister watching.

(There may be a good bit of creeping anthropomorphism in the tale as a whole, if not outright galloping mythomania; the report generally evokes a nice middle-class family down the block, more than anything else.) The troupe was observed most assiduously when it was eating; it partook of a menu of seeds, nuts, sweet roots, and grasses—never once did their lips touch meat—and brought into the camp large quantities of juniper and hemlock branches.

Through the years since Ostman's kidnapping, the record of Sasquatch behaviour has unfortunately not gotten much richer in detail, even though the number of actual sightings and firsthand print reports now approaches the four-figure mark.

The year Ostman was carried off is also the year that the first photographs of Sasquatch tracks were taken, in a section of land between Eureka and Cottonwood, California; these tracks measured a by-now rather modest fourteen and a half inches in length. There were also in that year reports of a night raid on a group of five prospectors near Mount St. Helens in Washington State. The men involved told of a band of man-apes heaving boulders onto the roof of their cabin and hurling themselves, screaming in fury, at the door. They were assumed to be retaliating for the potshots one of the miners had earlier taken at two of the animals in the woods; lawmen and reporters who subsequently visited the cabin found it badly damaged, and the ground around it covered with hundreds of enormous footprints. The area is now called Ape Canyon, and ever since 1924 it has been a prime area for both sightings and prints.

The year 1924 was the first year of such well-documented evidence concerning the Northwestern giant, and the first year of widespread interest due to this exposure; with these three separate instances spread across the Pacific Northwest in one year, the curtain on Sasquatch pulled open with a lurch and a vengeance.

The problem of Bigfoot observation is complicated by the fact that when a man watches an animal, the animal generally knows it, and man-observed traits can never be called perfectly natural ones. The behaviour pattern is always coloured by the need to fade back into the scenery at a certain point; whether surprised in the bush, on the road, or at the edge of a lake or stream, big feet at one point invariably beat either a hasty or lethargic retreat. No Jane Goodall or Dian Fossey has, since Ostman, been able to live and observe unnoticed in their midst, and until such a happy situation occurs there can be no real "scientific" assessment of their actual day-to-day conduct.

Nevertheless, observations have been made, and certain habits through the years cataloged, and from these emerges a rather makeshift list of what an eight-foot anthropoid has been seen doing or eating during the last fifty years. The first observations deal primarily with diet:

In 1933 a man named Cartright, along with a rock-hound friend, was enjoying his picnic lunch near the head of Pitt Lake in British Columbia, at an altitude of fifteen thousand feet. Also lunching, a quarter mile or so below the two men, was an animal with a "human face on a fur-clad body," feeding languidly on berries. The men, after discovering the large animal through their binoculars, continued watching until it finished its meal and strode out of sight. Carefully, they descended the rocky slope to where it had been standing and discovered its tracks in the earth, tracks they later reported as being quite distinct, and not at all bearish.

In 1941, twelve miles north of Agassiz, British Columbia, a woman named Jeannie Chapman ran out into her garden to see what was causing her daughter Rosie to carry on so. She saw an eight-to-ten-foot-tall creature, hairy, with a human face, approaching the house through the potato patch. She scooped up Rosie and herded her, along with her other children, out the front door and down to the river's edge. The animal clumped through the backyard, walked rings around the small house, and helped himself to a barrel of salt salmon in an outbuilding. After scattering fish heads and scales all over the shed floor, he proceeded down to the river, the Chapmans quickly proceeded back up to the house, and that, as far as they were concerned, was that. (Afterward, the tracks left in the potato patch and around the house were measured to be eighteen inches in length, eight in width, with a distance between the prints of roughly five feet.)

That the animal didn't actually chase the mother and her children down to the water's edge, but went down to drink only after sampling the salted fish, perhaps indicates a preference for fish flesh over people flesh, but is not very conclusive evidence of anything else. There are other reports of Bigfoot eating or carrying fish, usually during the autumn or winter months. One of these came nine years after the Chapman report, from British Columbia in the month of January, when the waters there contain substantially more sustenance than the earth. In midafternoon of January 7, 1970, the foreman of a logging crew unexpectedly came upon a seven-foot-tall, 250-pound creature on the road directly in front of him as he was heading back to the main camp. The animal, covered from the shoulders down with a reddish-brown coat of hair, ran across the road at his approach, in an upright position. The creature was, he reported, "a human-like animal" but

108

had "a monkey-type appearance." Having first thought from its size that it must be a bear, as he drew nearer he realized that it had a markedly simian cast to its features and stance, resembling, "a large, large monkey, and definitely not a bear." It had neither a bear's face ("more of a flat face, like a person or a monkey") nor gait when moving ("in the upright position, like a man would run").

As the foreman got within a hundred feet of the animal, it seemed to him that the expression on its face was one of either extreme terror or indignation—he couldn't be sure which. The proportion of arm length to body height was, he says, about equal to a man's, and the arms ended in what he describes as two hairy but humanlike hands. These hands were as hairy as the arms were hairy, and the right hand was holding a ten-inch fish. The muscles of the body were seemingly well developed, but due to the layer of four-inch hair that covered the flesh of the animal, these muscles were difficult to see. The one feature that did seem outstanding to the surprised logger was the animal's paunch—it seemed to be distended much in the fashion of that particularly hominid protuberance, the beer belly.

The foreman never actually saw the animal eating the fish it carried; after carefully regarding the man for a few moments, the erect creature moved across the road and up a dirt embankment at its side. It disappeared immediately into the thicket at the top of the bank, without ever putting fin to mouth. Nevertheless, it does seem a safe bet to make that during the fall and winter months, at least, fresh fish contribute substantially to the Sasquatch diet.

Three years earlier, in the autumn of 1967 in Marietta, Washington, a rash of Bigfoot sightings broke out during the fishing season. Upright creatures, estimated to be twice the size of an adult human, were seen sitting on log stumps at the edge of the Nooksack River, northwest of the San

109

Juan Archipelago. In the same area there were reports of such animals wading on the tide flats and generally making giant nuisances of themselves in the water. One gentleman out night-fishing felt a sizable tug at the end of his net and discovered a dripping Bigfoot at the end of it. The animal was pulling the net toward him, intent on helping himself to the fish it contained. After some shouting and a good bit of arm-flailing, the fisherman was able to enlist the help of four other fishermen nearby, and the collective glare of their spotlights induced the animal to relinquish his hold on the net and to abandon his not very sporting piscatory conduct.

Nooksack fishing slowed down for a while after this encounter, but after a week or so it resumed. An intrepid lady from south of Ferndale re-established her own fishing territory only to discover that Bigfoot had resumed his/hers as well; the animal appeared suddenly one night, looming out of the dark water within four feet of the side of the woman's boat. The appearance was made with a minimum of fuss and ripples—the fisherwoman as quickly and quietly abandoned her own sport. The next morning, thirteen-and-a-half-inch tracks were found along the shoreline and in the mud bank, tracks with quite distinct toe marks and heels, and with no discernible arch.

Aside from these reports of fish-eating, and the earlier reports of berry-eating, grass-, seed-, and sweet-root-gathering, there are other reports of interest concerning what our hairy giant puts in his mouth. In 1958 a construction worker on Bluff Creek Road—in the same area of Northern California where Roger Patterson was to make his Sasquatch movie nine years later—heard a considerable commotion outside the tool shack where he was sleeping and opened the door. He found an immense hairy beast standing directly in front of him. In a gesture of auto-

110

matic appeasement that undoubtedly surprised them both, the man picked up a chocolate bar from the table and handed it to the animal, acting out a terrified parody of trick-or-treat ritual. The animal took the candy bar and left.

Each area of the Pacific Northwest seems to have its own variety of Sasquatch fodder; along the Columbia River where Washington and British Columbia meet, the main attraction is garbage. The first report from that area came in the spring of 1969 from a woman who had seen two of the creatures poking through the trash cans along Williams Lake Road. That same April, two of the animals appeared at a point slightly farther west along the highway to Mrs. Betty Peterson, of Kettle Falls, Washington. Later in the year, in the autumn, tracks showed up in the mud near Swede's Pass, at a garbage dump near Bossburg, and near a boat-launching ramp beside Lake Roosevelt. These Bossburg tracks were cast in plaster and measured seventeen and a half inches in length, seven inches in width. (The Bossburg tracks have turned out to be some of the most interesting of all Bigfoot tracks, due to their excessively human appearance. The left foot is normal, but the right foot is what we would call a club foot. At its front the foot is twisted inward, giving the whole track a rather kidney-shaped aspect; there is a dislocation of bones along the outer rim, and the third toe has been squeezed out of its normal alignment.)

In the state of Washington Bossburg now enjoys a reputation for being prime Bigfoot territory, as the increasing number of sightings and print reports are filed and cataloged. René Dahinden, an ardent and knowledgeable Bigfoot stalker, counted 1,089 prints there after the 1969 sightings.

It ought to be noted here that chocolate bars and garbage

111

are not really what Nature intended as normal fare for any of her creatures—these food reports only serve to indicate how curious and adaptable Bigfoot appetites are. The most representative sighting of an animal eating is one made in 1955, in British Columbia by one William Roe, and it may turn out to be the one most indicative of natural eating habits in the wild. Mr. Roe met up with his giant while climbing Mica Mountain near Tete Jaune Cache, eighty miles west of Jasper, Alberta. He had the advantage of coming upon the animal as it grazed unawares; from a distance of seventy-five feet, he was able to watch it as it made its way through the low brush.

This, to the best of my recollection, is what the creature looked like and how it acted as it came across the clearing toward me. My first impression was of a huge man, about six feet tall, almost three feet wide, and probably weighing somewhere near three hundred pounds. It was covered from head to foot with dark brown silvertipped hair. But as it came closer I saw by its breasts that it was female.

As it slowly approached, Roe tried to make an accurate appraisal of its body size and shape, quickly deciding that it wasn't built at all like a female, with its frame as broad at the shoulder as at the hip. The animal's arms reached almost to its knees and were much shorter than a human's; its feet also were noticeably broader and thicker, about five inches wide at the front and tapering to much thinner heels. The soles of the feet, he was able to see, were of a grayish-brown colour.

It came to the edge of the bush I was hiding in, within twenty feet of me, and squatted down on its haunches. Reaching out its hands, it pulled the branches of bushes toward it and stripped the leaves with its teeth. Its lips

112

curled flexibly around the leaves as it ate. I was close enough to see that its teeth were white and even.

Finally, the animal saw Roe, and, still in a crouched position, moved backward three or four steps. Then it straightened to its full height, turned, and walked rapidly back in the direction of its original eating place in the shrubbery. As it neared this first site, it kept its head turned toward Roe; reaching it, the animal threw back its head and made a peculiar noise. Roe described this noise as being "half laugh and half language," sounding to him very like a kind of whinny. Roe waited for a time and then cautiously followed the animal's trail through the lodgepole pine and brush, curious as to what sort of stool evidence the animal might have left behind. What he was able to find bore out his idea that the large female was mainly vegetarian—there were, in the five sets of droppings he found, no traces of hair or shells of either mammal or insect. He also found what he assumed to be its sleeping place, a flattened-down area at the base of a large tree.

Though this individual was vegetarian through and through, there are other incidents where meat-eating has been observed, and it would be wrong to classify the Sasquatch as a thoroughly vegetarian species.

One such meat-eating incident took place in Oregon in 1967, during deer-hunting season. A young man working with a bulldozing crew clearing mountain roads chanced upon an entire family of animals as he rounded a bend in the mountain trail leading back to camp. On a high rock cliff some forty feet ahead of him, he saw two adults and a much younger animal between them, moving about good-sized rocks and smallish boulders, piling them one on top another. The man later described their faces as being "like a cat, without the ears," with noses too flat to be human.

113

One of the adults was much darker than the other, with hair a dirty brown colour. The lighter adult had a buckskin or fawn-coloured coat, and both had rounded shoulders and shaggy bodies. The darker animal had his longest hair on shoulders, neck, and head, with strands hanging down Hassidic fashion at the sides of the face, covering the ears. All three animals were equipped with virtually no neck at all—their heads seemed to sit squarely on their shoulders.

The animals were methodically stacking the rocks on top of one another, and with each upheaval poking underneath the rock's surface. The piles created by this activity were large, with each rock weighing eighty or ninety pounds, and there were sizable holes in the ledge where they had been removed. Describing them, the observer writes:

> He brought out what appeared to be a grass nest, possibly some stored hay that small rodents had stored there. He dug through that, and brought out the rodents. It seems they ate them. The rodents appeared to be in hibernation, or asleep or something. There were about six or eight rodents. The small animal, I noticed, only got one, but the others got two or three apiece. . . . They ate by just taking it in their hand and eating it as one of us would if we were eating a banana.

The three animals continued in their lifting and sorting of rocks—leaving rock piles of no fewer than three and no more than twenty pieces, spaced at intervals of between six and eight feet. They moved and ate methodically, in neither haste nor aggressive meat-eating fervor, sniffing tentatively at each of the rocks as they removed it. As soon as

114

they realized that they were being observed, they moved quietly away, leaving the rock piles where they stood.

The second incident of meat-eating and stone-piling was reported to me by Robert Morgan, a man who has headed up five major Pacific Northwest expeditions since 1968, and who is at this writing preparing for a sixth. (Morgan's tie-in with local and national conservation groups is an important one to mention—it insures Bigfoot protection from being gunned down in a machismic volley of game-hunting bullets, and promises instead the prospect of the animal being tracked, observed, and finally captured by the use of tranquilizer darts. The distinction between these methods is one that should be stressed by everyone involved in giant-hunting, everywhere that such hunting occurs.)

Morgan was out tracking in June 1970 in Skamania County, Washington, checking those streams and stream beds along which he believes Bigfoot travels, when he saw before him the red reflection of the animal's eyes, in the bush. It was well after nine o'clock in the evening, and rather than approach the animal in the darkness alone, Morgan hung up the electronic sound device he carried with him, on the limb of a tree. The device transmitted what Morgan hoped the animal would regard as non-threatening chord vibrations; he left the device attached to the bark of the tree and returned to his camp. The next evening he returned to find the sound device untouched, but with footprints pressed in the earth around it. Those prints that led away from the tree, however, became exceedingly difficult to follow after a few yards; it took Morgan all the next morning to track the prints a distance of one hundred yards. Near the place where he had first seen the eye reflections, Morgan found the fresh remains of a dead elk, a good-sized animal that had provided the larger one with a healthy meal.

In addition to the footprints and the elk carcass, there were piled up along the path stacks of small rocks that Morgan believes were left there by the animal as territorial marking devices or warning symbols. The freshness of the elk remains, the position of the footprints, and the piles of stones where none had been that first night, indicated to Morgan and his men that the animal was reacting to his presence, and to the presence of the humming machine, when it made the piles and when it made its retreat.

Zoologist and biologist Laymond Hardy has described to me a number of instances where, once a month or so, various primates reach out and pluck from the air or from a nearby branch bits of flesh and feathers, or flesh and scales, that serve as meaty complements to their usual vegetarian diet. That a giant primate should wolf down a bit of elk or yak or Hereford is therefore not an indication that the general diet is about to be given over to exclusively meaty pursuits. Any resourceful giant primate will certainly not turn up his catarrhine nose at a helping of well-seasoned flesh, should that helping fall into his copious lap, but this doesn't mean that meat-eating is his true and favoured style.

The Sasquatch stomach, so it would seem from those reports filed through 1974, is filled mainly with vegetable matter during the green months of spring and summer, but is supplemented with fish, red meat, and human refuse both in the green season and in the cold-weather months when the greenery has faded. As of April 1970, John Green's own Sasquatch file had recorded seven instances when berries were eaten, four sightings involving roots, two

116

apiece for tree shoots, grass, and deer, and four garbage sightings. Also reported were two instances of the animal overturning rocks in a stream and gobbling down whatever clung to their undersides. Once-apiece sightings included the consumption, or abduction, of goat, bear, sheep, chickens, mountain sheep, cows and horses, bones, rodents, grubs, clams, fish, salmon, evergreen buds, water plants, grapes, prunes, pancake flour, bacon and eggs, flour, milk, and a doughnut.

These reports and observations of the diet habits of the animal are significant reports, even if they are related to purely epiglottal behaviour. Such reports are the only possible means so far of tracing lineage, palatal and jawbone structure, dental set, and metabolic ratio, and will be most important when, and if, the time comes that Bigfoot is strapped down on some anthropologist-clinician's couch and made to say *Ah*. That time is not yet here, because the patient is still wandering about, and the doctor is more than a little embarrassed to set up his practice in the first place.

Bigfoot has, however, been seen doing a few other things out there in the Cascade bush and pine forest. In addition to filling his tummy with questionable nutrients, and building rock piles, he has also engaged in a bit of railing against the onslaught of what he must surely consider to be a wholly uncivilized civilization. His protests are infrequent, but thorough.

In 1963, in the same area where the chocolate bar incident occurred, there was another report of Sasquatch appearing at a construction site. This time, he showed up between two work gangs who were installing forty-eight-inch concrete culverts in the earth around Bluff Creek, California. Earlier in the installation process, there had been reports in the camp of these culverts being torn out

117

of their fresh beds and thrown into adjoining streams. On this occasion, though, the two work gangs returned to the truck carrying the main load of culverts at midday, and found that not only the culverts but the vehicle that carried them had been upended and flung over on its back. The truck was resting in the road with its load piled around it, wheels in the air, cab squashed beneath them. The two gangs between them were hard pressed to right the vehicle; that the overturning was the work of Bigfoot was a point of general agreement—the spate of fresh footprints in the gravel road later bore out this assumption.

This sort of conduct, a kind of monumental prankster-ism has been reported in other parts of the Pacific Northwest from time to time, most especially in and around the logging camps that house the anecdotal platforms for most real and imagined mountain activity. There is a great deal to be said concerning the metaphorical excesses of both mountain talk and mountain men—especially in a region with an extensive history of truth-stretching, such as the Pacific Coast. The Cascade Mountains themselves are said in some quarters to be the result of the melancholy excavation proceedings that were carried out to bury Babe the Blue Ox when he fell ill on Western shores, and Puget Sound itself is reported to have been created by similar grave-digging by Paul Bunyan, Big Swede, and all the rest of their outsize gang. Mr. Bunyan has, through the years of fable, thrown about quantities of lumber and dirt large enough to make Bigfoot's culvert-tossing and truck-upending activities appear as just so much playing with matchsticks.

The real Paul Bunyan who started the legend is now thought to be the large and muscular Paul Bunyon who gained fame in the Papineau Rebellion of 1837. At that

118

time a group of French Canadians, led by Bunyon, re-
volted against Queen Victoria in the first year of her Canadian reign. This Bunyon was a great fighter who later set
up a logging camp and became its "camp boss." By 1860,
his Canadian origins had been smoothed away, and he
enjoyed a reputation as a full-fledged American folk hero,
emerging as a figure larger than life and larger than any
one Canadian province. He was a giant among his fellow-
men, and his feats are now the main strands in a fabric of
folklore that stretches all the way from Maine to Vancouver
Island. That his success was so total and so immediate is a
matter to puzzle over; he is much younger in legendary
age than the Indians' hairy giant, and the one may have
benefited from the tales of the other. There may have been
a euphemistic kind of cultural overlapping concerning these
two, with the man Bunyon being cast in the role already
created by the quasi-man Sasquatch, in order to explain
what was, up until then, an inexplicable creature of North-
woods lore.

Whatever the relation of Bunyon to Bunyan to Bigfoot,
logging tales are still the repositories of certain backwoods
truths, unembellished by local legends. The facts behind the
legends are showing themselves more and more: footprints
in snow, footprints in mud, footprints on gravel roads, have
all been piling up through the last forty years with such
authentic frequency that to ignore them is impossible and
to blame them on familiar mountain animals is patently
unrealistic.

Wherever, or whomever, he may have been in the past,
the present Sasquatch is a sight to be reckoned with. He
or she may only appear for a few seconds, sucking on a
berry branch or downing a side of elk, but once he scampers
back into the Cascade underbrush, he leaves behind more

119

than just his furry fleeting image. It is by his feet, and his feet mainly, that we may be said to know him.

Lines of behavioural similarity among the three strains of mannish giant—in the Himalayas, the Andes, and throughout the Pacific Northwest—are not terribly easy to draw, particularly since they are cast against such diverse backdrops and partake of such different cultural interpretations. Observed behaviour, as we have earlier remarked, often has more to do with observer than subject, and in the Pacific Northwest, at least, there is certainly no predictable line that Bigfoot may be expected to take. (Robert Morgan has, however, been able to map an impressive network of migration routes, which he hopes will eventually prove their validity by revealing at least one migrating beast as he or she moves over the same path from seashore to mountain that others have moved over before.) The more specialized reports included on the preceding pages are not representative of the total body of sighting reports; most reports have the animal glimpsed for only a few seconds, doing absolutely nothing but standing erect or moving across a mountain road. Predictability, so far, has not been Bigfoot's long suit.

But when it comes to footprints, there is a bit more educated guessing to be done, and a bit more aligning of evidence, inasmuch as print casts cannot get up and walk away. In the Pacific Northwest Bigfoot prints most often fall into a category whose range extends from fourteen to eighteen inches in length. There are occasionally smaller tracks of twelve or thirteen inches that turn up, and every once in a while a really gargantuan print, but the main inch range is between fourteen and eighteen. The largest print so far is twenty-four inches long.

120

In the Himalayas the range of steps is slightly greater, with a six-inch measurement at the bottom rung and twenty-four inches again at the top. Eric Shipton's 1951 tracks from the Menlung Glacier, the cleanest of the Asian prints, measure thirteen inches in length and eight inches in width. As a general rule, Himalayan footprints are from three to four inches shorter than their North American counterparts, and are found most often at altitudes greater than fifteen thousand feet. Whereas these Yeti prints are pressed in layers of snow along the mountain slopes and fields, Sasquatch prints are found most often in mud, dust, sand, or in river silt, and only rarely in snow.

South American prints, when they have been found, have been discovered almost always along the banks of rivers—as in the case of the recent prints found in Ecuador, stamped into the mud at an altitude of thirteen thousand feet, and as with the twenty-one-inch tracks from along the banks of the River Araguaya. It is hardly surprising that the most populous areas of giant territory yield up the greatest number of prints, and that South America's remote jungles and rivers trail behind North America and Asia when it comes to print sightings and reports. Whereas both the Tibetan Government and the good burghers of Skamania County in northern Washington have levied fines and licenses concerning the pursuit of their native Brobdingnagians, Colombians and Brazilians and Ecuadorians have introduced no such strictures against giant-stalking and killing. As the Amazon highway cuts its way through the forest, though, there will almost certainly be more attention focused on the wildlife disturbed by the intrusion. With every new flotilla of hover-craft that plies the Xingu and the Negro, opportunities increase for giant sightings and print discoveries along their banks.

The way the record now stands, the American Pacific

Northwest is fuzzy head and muscular shoulders above both rival areas when it comes to track reports, and the sightings and plaster casts continue to pour in. That faked prints may more easily be laid down by pranksters in their Spokane suburbs than by Sherpas in Sikkim or by Wauras in Cachimbo is a point beyond disputing. In the summer months Northern Californians and Washingtonians are not always shod, and the bigger folk among them have probably left any number of intriguing trails through the bush. (Regarding length, human feet have been known to extend as far in dimension as the great majority of Yeti and Sasquatch tracks do—but only in rare instances do they pass the fifteen-inch mark. One remarkable example of the *Homo sapiens* exception was Eddie Carmel, the nine-foot-tall Jewish giant of Diane Arbus's famous photograph. His feet were sixteen and a half inches long, a full two inches better than the subject of Roger Patterson's 1967 film from Bluff Creek.)

People tracks and Bigfoot tracks are not the same thing, though, for reasons other than length alone. There is the matter of weight distribution and over-all width, plus the particular structure of particular toes. (This question of toe structure also separates them without question from any lingering accusations of beardom that might still cling to them. Bear tracks, as both Bernard Heuvelmans and John Napier have pointed out, are really rather like mirror reflections of hominid tracks; when the ursine glass is held up, it reveals a big toe where our primate little toe should be, and a little toe in place of our big one. Bears, unless they were so pigeon-toed as to be ridiculous, could hardly swing their left paw into and across the right lane, and the right across the left lane, as a general rule of walking, and still manage to get from one place to another. Bears, for once and for all, are out.)

122

The Bigfoot toes in the configuration which has been dubbed by Professor Napier as the "hourglass print" have a tendency to line up neatly, in the sand, mud, or river silt, like a row of corn kernels on a cob. The big toe in this type of hourglass print—which comes more often from the state of California than from either Oregon or Washington or B.C.—is not as large proportionately as the big toe in human prints. Also, the line is generally corncob straight, whereas human toes tend to curve subtly and present a ridge behind the toes which pushes forward with each increase in toe size, from little to big. Also, the toes themselves are longer and more "apelike" than human toes. These hourglass prints also show a difference from human prints in their heaviest point of weight distribution. Whereas man walks with a thrusting motion from his big toe and the inside of his foot, the giant prints of California thrust forward from the outside of the foot, and thereby evoke a very different pattern of walking from the one popularly used by humankind.

More akin to our own barefoot human tracks are those that a large number of Big feet have left behind in Oregon, Washington, and over the border in British Columbia. These differ from those with the hourglass configuration— from Bluff Creek and other areas in Northern California— in that structurally they resemble actual human tracks more closely. The feet of this northern track-maker have a much greater flatness than do his relations to the south, and show a nearly undifferentiated sole plane, a toe ridge noticeably more oblique, and an emphasis of thrust that produces a different stride entirely. The big toe is a proper big toe, and each of the smaller toes has a discretely human structural character; the whole foot is proportionately flatter and longer and isn't nipped in at its middle the way the hourglass type is. Although there are some further refine-

123

ments in both these two types of prints, tracks in the Pacific Northwest have been divided up rather neatly as belonging to either one type or the other.

(It would be nice, and wonderfully neat, if the two types of observed Himalayan giant also left tracks as obligingly diverse as the Cascade wanderer has, and if the Brazilian and Ecuadorian varieties were split on a similar hair. Such, alas, is not the case, and we are left with neatly articulated differences concerning the tracks of one species range, but with no such convenient subdivisions in the ranges of the other two. Furthermore, this pattern of North American tracks may well have to be expanded and revised due to the prints coming in from areas well out of the Cascade Range. Recently, I have seen prints in the central Florida ranchland bordering the Everglades which complicate the track record even more. These showed up at Debbie and Michael Polenek's ranch, where a yearling horse had been killed and ripped apart, and a calf beheaded and stripped of its flesh. The tracks found near the yearling were shorter than Sasquatch tracks are generally, and showed three toes rather than five, but they had a similar heel impression and structure. The killings occurred in the same area of the state where local police officer Joe Simboli reported an encounter with a seven-foot-tall hair-covered creature twelve months before, and where there have since been recurring reports of ape appearances, and ranch animals being attacked. There is now increasing evidence supporting the idea that Bigfoot is not a purely northwestern phenomenon; sightings have been made and tracks reported in thirteen states, with prints extending all the way south to the Florida Keys.)

Back in the land of Sasquatch, there is one bit of additional categorizing that may be done—between sightings concerning the actual animal and sightings concerning only his tracks. Actual flesh-and-blood sightings have been made more frequently in the United States, while footprints have been more numerous across the border in Canada. This fact, like many facts concerning the elusive animal, probably has less to do with the character of Sasquatch than with the character of the terrain in the two countries, and with the difference in population density, hardness of gravel, and tightness of lip between the two areas. But it does reflect the compiled data over the years and may turn out to be a significant behaviour index, after all.

Why the divisions between print types occur, and why the animal is seemingly more egregious in Canada than in the United States, is, really, anybody's guess. To pretend that we are able to stick a Sasquatch slide under the microscope of modern science when there is not yet any such slide to stick, and deduce specific causes when we are still unsure of any effect—this, of course, is the worst kind of ontological frippery.

We are all amateurs when it comes to giant-stalking, but this very amateurism may be the best of all qualifications when it comes time to finally put into perspective the Bigfoot past, present, and future. There is something compelling about the urgency with which he is now being pursued —compelling and fascinating and indicative of an interest that transcends the hunter's interest in his game or the ethologist's interest in his subject. Probably, in varying degrees, we are all fascinated by the prospect of encountering someone or something so very unlike ourselves, while being at the same time so very like us.

Probably a kind of cross-species mesmerism is going on, with *Homo sapiens* playing alternately the role of mongoose and cobra, and with Bigfoot cast as both villain and hero, angel and beast.

Probably, we are all Fay Wray.

125

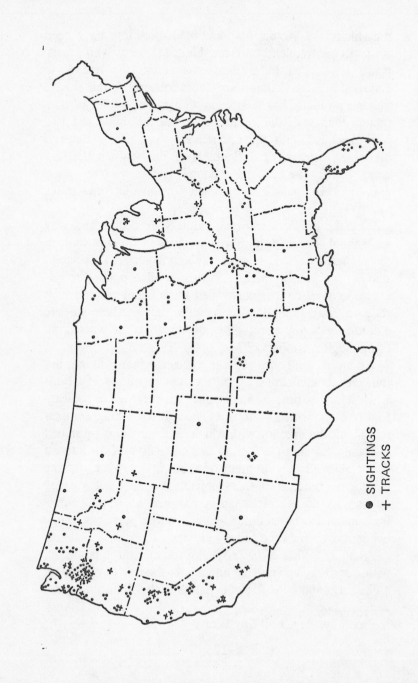

● SIGHTINGS
+ TRACKS

Sasquatch, undoubtedly, lives. As of this writing, the most recent photograph taken is from April 1973 and shows a somewhat pensive animal propped up against a rock looking out across a rocky cliff. This was before the fiery cataclysm that spread through the Pacific Northwest later that year, but the odds are that at least some of the animals were able to survive even that disaster. Whatever the number of his fellows, and however diminished his range may have become because of the damage, the Cascade giant, like the coelacanth in his element and the arctic tern in his, must surely still endure. Perhaps, with his vegetarian range lessened and with the inevitable encroachment of civilization on the little territory he has left—with these things imminent, the Bigfoot diet will come to include more and more protein and less and less starch and carbohydrates. As Ronald McDonald and Colonel Sanders approach the wilderness—turning forest to suburb and redwood to plastic—there would be a sweet irony in our anthropoid cousin's eating habits approaching our own. I confess to a dream of the now remote and benign giant families growing canny and aggressive on Big Macs and monosodium glutamate. Multiplying at a faster and faster rate, they would turn Arthur Treacher fish cakes and Roy Rogers beef sandwiches to their advantage along the way, growing more dominant with each bite. At last, of course, they would develop a taste for human flesh. As the last scene in my dream tableau I see a smiling Sasquatch franchisee serving finger-lickin' good buckets of Businessman's Buttock and Breast of Starlet to a line of furry teen-agers.

A delicious fancy, I think, and one that turns inside out the pattern of what has been for so many millions of years respectable hominid behaviour:

Sometime, long, long ago when big apes and small apes moved down and away from the trees, the taste for flesh

127

was acquired by one group of animals but was not picked up by the other. Who or what was it, I wonder, that whispered into the ear of the leader of the Cain tribe suggesting that the flesh of the Abel tribe might be tasty indeed? Once having eaten and enjoyed flesh, is it not possible that the innovators realized their advantage, and did the best they could to keep their outsize primate brothers from thinking how nice it would be to in turn eat them?

The killing of giants must have begun very early on in primate history; in addition to implicit self-defense, there is the backsliding question, the matter of keeping one's developing human line away from a more "bestial" giant line that might pull all the refinements backward again. The meat-eating man-ape perhaps for these reasons reflected to himself that exerting dominion "over the fish of the sea, and over the fowl of the air, and over the cattle, and over all the earth, and over every creeping thing" meant knocking the brains out of any alternate species of still-grazing primates. These primate "cattle," these borderline human-kine, were, then, definitely to be subdued.

It is my belief that these early human-kine, these giant graminivorous primates still nibbling at the roots and leaves, are the same giants referred to in many of the world myths, evoking a collective race-memory that goes back to the beginnings of all myths and legends, the first tentative rustlings of what we call knowledge. A corollary to this belief is that our present-day giants on the three continents of Asia, North America, and South America are stragglers from this early giant parade who have lost themselves along the sidelines.

There are those people among us who think that the Genesis myth is not merely a myth and that the Garden of Eden was not quite what it is pictured to be on our pastel-coloured Sunday school cards. There are also many who believe that the first garden, as mentioned in Chapter Three, was an earth chakra located high in the Andes, and later the Himalayas, in which earthly experiments were carried out by nonearthly "gods" or astronauts from another part of the universe. There are, finally, others who maintain that these "gods" were very large indeed, and that they did a good bit of experimenting with size and diet, and rather a lot of solar energy research, attempting to create facsimiles of themselves on this planet because their own planet had become nearly uninhabitable.

(If our own early astronauts had moved out of their suborbital pattern and bounced through time and space to Venus or Saturn, the cross-colonization of space would have taken a curious turn; in 1961 the first astronaut was named Ham and was a chimpanzee.)

The existence of giants alive today proves nothing about yesterday's gardens, gods, or chakras. It is impossible to deduce from a footprint in the sand or a bit of jerky celluloid the reasons for floods, arks, or human evolution, or to see the rings of Saturn in the calcium layers of a giant lower molar.

But it is possible to think of these things, and to speculate how they touch each other, and to entertain theories as theories only. It is possible, with the footprints before us, and the lower molars in hand, to set out on a short journey where we may or may not be able to gather some facts. Let us, then, close our books and go back into the Garden.

Five

There is a popular legend from ancient China that concerns a great Stone-Ape, his creation and adventures. The story goes that, back in the beginnings of time, there was an island lying in the middle of the Great Eastern Sea called the Mountain of Flowers and Fruit. There was on this island a huge rock which had absorbed all the hidden power of heaven and earth and of the sun and moon and planets, ever since the beginnings of the world. This rock one day burst open, and out of it emerged the Stone-Ape, as he was called, and this creature immediately bowed to all four corners of the earth and went about the business of exploring the island and creating his own tribe. After a time, this Stone-Ape led other apes of his tribe through the island's magic waterfall and into the Cave of Heaven behind it. In this cave, over a period of three hundred years, the Stone-Ape was able to learn seventy-two magic arts and occult practices which enabled him to one day steal a powerful iron rod from the Serpent King (or Dragon King) and journey to the Underworld of the Dead. There he demanded to see the Book of Life and Death, where all names and life-spans were recorded. Ripping out his own name and span from where it was written, he declared himself immortal. Another time, he stole into the home of the god Lao-tzu, in the Western Heaven, and drank up the

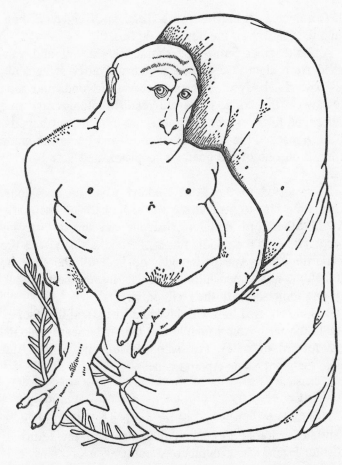

liquid from the five gourds containing the Elixir of Long Life. Again he declared that he would live forever.

Then he descended to earth.

On earth he became so powerful that finally the Lord Buddha appeared to him, with the intention of humbling him and taming him.

Buddha said to him, "I understand that you turn somersaults over the white clouds of Heaven, and that each of

131

these somersaults carries you a distance of eighteen thousand li. Please show me how you do this."

And the great Stone-Ape shouted his assent and vanished from sight. He turned somersault after somersault, his powerful body creating a mighty whirlwind, and soon he saw on the horizon five tall reddish columns towering taller than his own great height. He said to himself happily, "I have reached the end of the world!" and he hastily made a mark on one of the columns as proof, and flew back to earth.

He reported to the Lord Buddha what he had done, instructing him to go and see where he had left his mark. At this, the Lord Buddha cried out and thrust forth his hand. There on the middle finger of Buddha's hand was the mark the Stone-Ape had made; he had, all the time he thought he was somersaulting through Heaven, been merely turning somersaults in the god's palm.

Buddha covered him with his hand, then, and imprisoned him in the earth from which he'd come, underneath a mountain formed of water, fire, wood, earth, and metal. And there the great Stone-Ape remained.

This fable of hubris and its reward is a tale that might be translated without too much difficulty into the terms of existing legends in a number of cultures; it contains in it certain motifs that appear over and over again in fables ages and continents apart. The garden on the mountain, along with the secret cave and the surrounding water are settings not at all infrequent, everywhere from Macronesia to Barnegat Bay. They crop up in current stories and fables, as well as in ancient ones, but the interesting thing is how many cultures give the same creatures the same attributes or myth functions. Whether the serpent from the Garden of

Eden, the Midgard serpent, or the oriental serpent-dragon, the great snake is generally seen as the keeper of some peculiar power, wisdom, or magic. King Kong wrestling with the serpent is a direct confrontation, while Eve giving her ear to his insidious whisperings is an example of a more subtle approach to the conflict; the iron rod that the Stone-Ape takes to enter the chamber where the Book of Life and Death is kept and the apple that Adam eats to understand what the Tree of Knowledge is all about are both potent myth objects.

But the figure of the great ape as hero is one not nearly so common as the setting he is placed in. (It is a neat coincidence that this ancient folktale comes from the land where ancient giant bones were first uncovered, and where Dr. von Koenigswald came up with his first *Gigantopithecus* molars in the 1930s.) The giant apes that occur in the folklore of other cultures are never so presumptuous as the Stone-Ape, and most outsize bipedal creatures are not referred to as apes, but simply as giants. Granted, a number of these are covered over with hair or hairy pelts, but history and prehistory are neither as thorough nor as precise as the Yeti and Sasquatch watchers of this century.

Still, there are details.

In the epic of Gilgamesh, that story of ancient Sumeria which was found on twelve stone tablets at the end of the nineteenth century, there is a character named Enkidu. He was created by the goddess of heaven, Aruru, as a rival for the heroic Gilgamesh, and his description suggests a creature with a remarkably *Gigantopithecine* cast. He was half-man and half-beast, his body was covered all over with coarse hair, "he knew neither kin nor country," and "with the gazelles he ate the grass." He is described grazing and also drinking from a pond with cattle. This large beast was tempted to mate with an attractive princess, partially divine, and to so become more amenable to civilization and to Gilgamesh himself. This hominid hanky-panky had its effect; Enkidu and Gilgamesh became fast friends and went buzzing around the universe together. There are in the epic a number of parallels with the Genesis story; to my mind the most noticeable is the animal-human melding, which is carried out over a period of six days and six nights. Genesis 6:2 relates:

> That the sons of God saw the daughters of men that they were fair; and they took them wives of all which they chose

and then continues on, in the third and fourth verse with the outcome of this sort of comingling.

> And the LORD said, My spirit shall not always strive with man, for that he also is flesh: yet his days shall be an hundred and twenty years.
> There were giants in the earth in those days; and also after that, when the sons of God came in unto the daughters of men, and they bare children to them, the same became mighty men which were of old, men of renown.

Whatever else may be deduced from these entries, and from the parallel Gilgamesh chronicle, it is certain that there were two different types of primate, having at each other sexually. Whether Enkidu (or Ea-bani) was one of the "giants in the earth" (or Nephilim) and whether he was the father of one of these "mighty men" is less clear. The question of nomenclature is a nubby one, but the salient points that may be extracted from these two accounts are that giants were mating with nongiant types at one point (and not making the gods, or God, very happy by doing so), and that such mixed mating produced outsize results, who were to be giants in their own right.

That the figure of God, either as alien astronaut or Buddhic overseer, should have expressed disapproval of a practice that was common in all other genera, including the best of primate families, indicates that he had more ambitious plans for his latest creation than any back-breeding with previous open-air suite-mates would allow.

Apparently, giants, for one reason or another, did not contain the stuff of which heavenly aspirations and dreams were made, and the God-figure decided that anyone of the elite (Sons of God) who was caught dallying with a lady of the pre-enlightenment evening (Daughters of Men) would be drummed out of the ongoing *sapiens* corps. (The Hebrew verb *naphal,* which means "to fall" or "to fall upon," is the base for the term "nephilim," which is the Old Testament translation for giants; it may have been that "fallen" men and "fallen" women were those whose parent, on one side, forgot what he or she was supposed to be about long enough to mate with less thoroughly evolved sistern or brethren, on the other. Or the nephilim might have fallen from the trees, or from the skies. . . .)

Later in the Genesis story, there is an incident absent in the King James Version but present in various scattered references in the Hebrew Midrash which concerns a giant named Og who is said to have ridden on the ark during the forty days and forty nights of the downpour that caused Noah's flood. In order to be included in the ship's company, so the story goes, it was agreed that he would, once the ark reached dry land and the waters receded, become a servant of Noah and of all his descendants. He was the only one, of what was a widespread race of giants, who was able to survive the flood. According to the Midrash, all the others of

135

the ancient tribe perished, along with all men but Noah and his three sons, Shem, Ham, and Japheth.

In the ancient and heroic sagas of Norse mythology, the Eddas, there are numerous stories of giants and of couplings between the daughters of these primitive creatures and the sons of the gods. The first of these occurs at the early beginnings of time and life on earth, when the first giant, Ymir, is created out of the frozen gases and mists of the region called Niflheim, the place of primeval frost. (The similarity of the words "Niflheim" and "Nephilim" is immediately apparent; that these two words should occur at opposite ends of the world in such disparate languages and cultures, but with such similar meanings, is both curious and astounding.) This Niflheim giant, Ymir, licked by the flame of Muspelheim, produces from somewhere between his third rib and his armpit two more giants, who set about populating the hardening mists. Shortly thereafter, licked into being by a great nameless ice cow, who materializes beside Ymir (like a Norse Babe beside Paul Bunyan), something smaller and more comely than these ragged giants emerges from the swirling mists. Whether a man or a god, the fair, straight, and handsome being mates with a daughter of the hairy, crooked, and unkempt giant tribe and produces three heirs. During this early age, a type of monumental miscegenation takes place with this, the first godlike creature, impregnating the daughter of one of the first of the race of shaggy giants, or *jotuns,* and founding thereby a whole cosmogony. The three offspring are the three great gods of the Aesir pantheon: Odin, Hoenir, and Lodur.

Before these three young gods can go about shaping the rest of the world around them, the Edda records that it

136

is necessary for them to commit genetic patricide; they are required to kill their own maternal ancestor, the by-now ancient jotun, Ymir. They dispatch him quickly but with such attendant commotion that a good many other giants are killed as well. In fact, all those jotuns who have not mated with the god race are drowned in the great flood that Ymir's death precipitates. All, except for one young giant and his mate who somehow manage to survive the flood by floating through it on an ice floe to a land where the gods do not wish to follow them. The land they reach is dark and very cold, not far from the shores of a great sea and is called, then, Jotunheim.

After the flood, the three god-brothers set up their own home on the highest mountain, and create the first true-man, Ask, and the first true-woman, Embla. These two breed, and the offspring of their union are described as being quite unattractive, with rough skin covered by furs, with gnarled joints, and downcast eyes. They are kept fenced in by the gods, away from any threat of marauding giants, and gradually, generation by generation, their appearance improves, and their children's children begin to look more and more like those beings the gods intended. The giants, though, are always a threat, and the gods and their own children are forever keeping them at bay, away from the human colony.

Throughout both the Prose Edda and the Poetic Edda, there is a continuous battle between the gods in their mountain home and the angry jotuns. Thor, son of Odin, is often pictured doing battle with them, his iron mitt extended, his "magic hammer" in hand, and his "magic belt" cinched round his waist. (The hammer is a kind of red-hot boomerang that sounds rather like something William Shatner or Leonard Nimoy would have used—it shot out of Thor's hand, immobilized or crushed its victim, then

137

bounced back to his protectively mitted hand. The belt had to do with increasing his weight and his mobility.)

However, one giant did manage to find his way to the top of the Asgard mountain, and it was he who finally caused the destruction of the Aesir, or god race. This was the jotun Loki, who, fairer than his fellow jotuns, won the heart of Odin and was spared by him. The two, when they were young, had opened the veins of their arms and become blood brothers. Loki was thereafter made a kind of honorary god and was welcomed to Asgard by Odin's real brothers and by the rest of the god community. Like Enkidu in the Gilgamesh epic, Loki was given a goddess-wife, with whom he mated, and was made to leave behind in Jotunheim his hairy and unattractive giantess mate, Angerboda.

The giant Loki never made the grade in Asgard, reverting over and over again to the nastiest kind of bitchery and the most treacherous sort of jealous raillery. To list all the evil and malevolent acts that Loki engaged in during his time among the gods is here impossible; he was a fetching psychopath whose size and beauty enabled him to get away with innumerable acts of cruelty and malice. Having god blood in his veins, thanks to Odin's impetuosity, he remained immune to Thor's hammer and was generally forgiven all his misdemeanors against men and gods. As pictured in the Eddas, he seems to represent unbridled emotion and physicality, contrasted against the calmer and more benign rationalism of the Aesir; his moral code is nonexistent—all that ever prompts him to action is an attack on his emotions or his closest friends. He is altogether incorrigible, a thoroughly nasty fellow. When he at last reveals himself, in a fit of petulance, to be responsible for the death of the gentle god Balder, the family of gods can no longer excuse or forgive him—they tie him to a rock and drape over him a poisonous snake, whose venom is destined to drip down on him for all eternity.

With the death of Loki, there follows a period of total chaos; all heaven and earth are thrown into an uproar as Odin is swallowed up by the monster Fenris, and Thor, after a great battle, is killed by the "poisonous breath" of the Midgard serpent. All the earlier imprecations against giants, and the warnings against mingling with them, seem borne out by this final orgy of destruction. The inhabitants of the savage earth are seen rising up and destroying the Aesir colonizers, in the process avenging the death of the first giant, Ymir. And finally, just as the human race was created after the death of the early giants in the great brine-flood, so a new race rises up after this destruction of the gods. This pattern of regeneration after megacide is one not uncommon in cultures other than the culture of the Eddas.

The ancient Orphic cult of Greece has its tale of regeneration after the killing of giants, drawing from the myth of Dionysus and of his capture by the giant Titans. Captured by the Titans, the god Dionysus was killed by them and his body eaten by the giant tribe. This so angered Zeus that he in turn killed them with his lightning and thunderbolts; from the ashes of the dead giants—which contained as well the remnants of the divine Dionysus—there emerged a new race, the race of man. (Orphism added the concept of the transmigration of souls to the original Titan story, and later Pythagoras built a whole ethical system around it, incorporating the idea of the perfectability of man through reincarnation. Depending on behaviour, the soul of a man returned to the gods or was punished for its sins on earth by a return to the earth.)

That living or dead giants were a necessary condition to the creation of mankind is a recurring mythical fact. Whether they were called jotuns, Titans, or, as in Tiahua-

nacu, huaris, or gathered in the biblical tribes of the Nephilim, the Gibborim, or Rephaim, they existed as a race
apart from the developing race of man, often doing battle
with it, but more often being the primal stock out of which
the human race, through mutations, cataclysms, and lightning bolts, emerged.

Wherever it was that these ancient giants were coming
from, and whatever reason they had for the trip, they left
their image in the lore of a large number of early cultures
as they passed over the face of the earth.

In the legends of the Cochiti people of the Pueblo nation
in New Mexico, there is a story of a sacred place, a cave
where one such shaggy giant was shut up long ago. He was
kept away from mankind, imprisoned in a rock cave called
the *gashpeta*. This sacred place, this gashpeta, is where the
giant remains are still said to exist, sealed up and removed
from any further contact with the human race. In the same
state, two hundred miles west of Fort Sumter, there is another Pueblo village called the Abo settlement, where on
the stone walls of its open caves are painted in earth colours
representations of the feet of man, and of other animals of
the region. Sticking out like two sore thumbs on a platter of
ladyfingers are prints nearly three times as large as human
prints, but with the same outline of sole and toe and the
same stylized structure.

This giant cave and these giant tracks are indications that
Sasquatch, or someone very like him, may have once passed
this way. Any giant tribe on its way from the Andes to the
Cascades might well have trod this same dry New Mexico
land, when it was considerably wetter, either lingering only
long enough to press its footprints into the earth or else
choosing to stay for generations in and around the area of
the gashpeta cave.

If thousands of years ago there occurred a northward

140

migration that moved the animals up and through the present United States, it is probable that the animals followed the rivers as well as the mountains, in the same way that they are thought to follow the rivers today. Along the Mississippi, and all across the arc of the Gulf Coast, there are still legends of the dangerous loup-garou, a Cajun animal, half-man and half-beast, who is reputed to frighten little children and make off with cattle and sheep during the dead of night. Throughout the countries of Europe and Asia, as well as in America, the stories and characteristics which accrue to wer-wolves and their fearsome ilk are, perhaps, related to apes in wolves' clothing, rather more than to wolves in man's.

The smoke of such beastly legends does not automatically vouch for any bonfire of fact beneath it; still, the great *Urp-tier* that lies at the heart of these various beast fables could as easily be the very real great-great-grandson of *Gigantopithecus* as the son of Dracula or the daughter of Frankenstein. However, it may be slightly too extravagant to attempt to pin all the tails of Loki, Enkidu, Og, the Chinese Stone-Ape, and the members of the Genesis giant tribes on the *Gigantopithecus*-donkey, while at the same time colouring in his image on the wall to resemble Lon Chaney, Jr.

I will not press the wer-wolf point.

It is a long time since Genesis. No matter precisely where nor exactly when the events in that book took place, they do record the earliest beginnings of things as far as man is concerned. And *Gigantopithecus,* or someone very like him, appears in the record of those early beginnings as an earnest and inelegant doppelgänger to primal man, very much in evidence at the time of those early goings-on.

The Homo-Adam, barely sapient and still self-conscious about standing erect, was flanked by his primate cousins and brothers standing in much larger numbers around him than they stand in today. Perhaps he is to be excused for his treatment of them as he grew—for the ruthlessness of the aggressive success that he achieved, climbing up over their less efficient bodies and bones. Perhaps he was a kind of giant himself.

Pierre Teilhard de Chardin cites considerable variation within this early hominid clump, up to and through the Quaternary period, fifty thousand years ago. In *The Appearance of Man* he writes:

> Thanks to the efforts of the Geological Survey of Bandung (and more especially to Dr. von Koenigswald), at least three different Pithecanthropians are today identified in the lower (Trinil) or even basal Quaternary (Djetis) of Java, *Pithecanthropus erectus, Pithecanthropus robustus, Meganthropus paleojavaniens,* and perhaps yet a fourth form. These, added to *Gigantopithecus* of southern China, and to the *Sinanthropus* of Peking, represents, on the edges of the Pacific, a half-dozen closely associated characteristic forms.

After considerable work in co-operation with Dr. Pei Wen-chung, Teilhard is able to conclude that

> Forty or fifty thousand years ago, that is to say at the time when the loess was beginning to be deposited in China, humanity—though far from having yet attained its present anatomical stage—formed an extremely complex assembly in which the anthropological types were much more divided than in our own modern races.

142

Teilhard is not willing to go so far as a number of his associates, those who would bring *Gigantopithecus* into the circle of true humanity, but the anthropoid background he describes is rich indeed, quite rich enough to admit all manner of possible combinations.

Franz Weidenreich, in *Apes, Giants, and Man,* is less reluctant to go all the way with this early giant, and gives him full human honours:

> The giants and near giant-forms are connected with the normal-sized early types. The human line, especially the most primitive group, has been considerably extended by these new discoveries and by the more correct interpretation of *Pithecanthropus robustus* as a form of intermediate between the normal-sized and the giants. I believe that all these forms have to be ranged in the human line and that the human line leads to giants, the farther back it is traced. In other words, the giants may be directly ancestral to man.

Because of such common ancestry, Weidenreich has even suggested that the Asian giant has been given the wrong classification. "Indeed," he writes, *"Gigantopithecus* has been mis-named; it should have been called *Gigantanthropus."*

-Anthropus or *-pithecus,* whether giant preman or colossal ape, the creature is not merely an evolutionary footnote, but something inestimably more important. It is truly amazing that he can exist, like the coelacanth, both as an ancient fossil and as a present flesh-and-blood figure moving his enormous weight across the world landscape.

Less amazing, and considerably less wonderful, is the fact that, apparently, we killed him then, and, most probably, we will kill him now.

143

In a November 1973 issue of the New York *Times* there was an article titled "Yeti-like Monster Gives Staid Town in Illinois a Fright." The article concerns a town, Murphysboro, Illinois, some twenty miles east of the Mississippi, where, all through the summer and autumn of 1973, a giant hairy creature, eight feet tall and light-coloured, has been glimpsed roaming the shores of the Big Muddy River, frightening the townspeople and leaving his tracks pressed in the mud. Those people who have seen him, the Cheryl Rays and Otis Norrises and Wes Lavandars, describe an animal weighing three hundred to four hundred pounds, who appears to be more curious than menacing. He has been observed standing quietly in the brush along the riverbank, with his head cocked, watching tethered carnival ponies or slowly approaching parked cars. Occasionally, he has been heard wailing in the marshes, emitting a cry which Murphysboro people liken to the sound of a greatly amplified eagle shriek.

The town is frightened and alarmed, as towns in the Pacific Northwest are frightened after a sighting, or as horse-breeders in Florida or yak-herders in Tibet are alarmed when fresh prints are discovered there. The sound of the creature is everywhere reported as terrifying, capable of whitening hair—as in Turolla's Andean cave—while the sight of the animal is one that sends most people fleeing. But the present-day Sasquatch, Bigfoot, or Yeti is not a monster—he is only a giant, and much less a murderer than man himself these days. He has kept alive by keeping apart, and that, it would seem, is the only way he can continue to survive at all.

The Murphysboro article ends with teen-ager Randy Creath, who confronted the creature from a distance of fifteen feet, reflecting: "I know it's out there. It would be fascinating to see it again and study it. But, you know, I kind of hope he doesn't come back. With everyone running around with guns and sticks, he really wouldn't have much of a chance, would he?"

The answer to this question is, of course, No, he wouldn't.

Émile Zola wrote, "The fate of animals is indissolubly connected with the fate of men."

There is no better example of this connection than the bond which still ties our anthropoid line together and links us with our present living giant. It is obvious that he has a great deal to tell us. And yet, Zola's simple truth goes unheeded, its urgency ignored, its implicit warning disregarded. Man will ignore the bond, and so thoroughly prune the earth's shrubbery and pollute its rivers and marshes that no outsize species will long be able to hide from his shears and his defoliants and his guns. The giant anthropoids of the Americas and Asia have no business, really, hanging on as long as they have in the face of such energy—by all rights and reckoning, they should have died out, well before the last ice age, leaving only their frozen bones and calcified teeth to speak to us today.

But the thing is, they don't know that.

The coelacanth in the African ocean swims silently past the knots of wriggling sea vipers where they breed, leering in the sludge, unaware that in their yellow eyes he is just another load of warmish waste, a chunk of finned garbage they'd like to get their teeth into, a bit of flesh they'd love to wind their blue-gray tongues around.

The Sasquatch in the California night wanders through the Cascade foothills, striding beneath the stars, past turquoise and pink motels, concrete and asphalt trailer parks,

145

where hunters and campers and adventurous men lie sleeping. He is unaware of the men's dreams, of their images of shaggy giant trophies on their walls, of furry skins hung over their portable TVs.

Night and water and green leaves are what interest these two animals, fish and mammal—not vipers or men, or the space between the stars. They continue on their separate ways, unaware, as they have remained unaware for hundreds and thousands and millions of years, that they do not exist at all.

Works Cited

Chapter One

Heuvelmans, Bernard. *On the Track of Unknown Animals.* New York: Hill & Wang, 1965.

Simons, E. L., and Ettel, P. E. "Gigantopithecus." *Scientific American* 222 (January 1970).

Waddell, L. A. *Among the Himalayas.* London: Constable, 1899.

Weidenreich, Franz. *Apes, Giants, and Man.* Chicago: University of Chicago Press, 1946.

Chapter Two

Carpenter, Frances. *South American Wonder Tales.* Chicago: Follet, 1969.

Hartt, C. Frederick. *Mytho do Curupira.* 1879.

Heuvelmans, Bernard. *On the Track of Unknown Animals.*

Homet, Marcel. *Sons of the Sun.* London: Neville Spearman, 1963.

Marsh, Richard Oglesby. *White Indians of Darien.* New York: Putnam, 1934.

Marshall, N. B. *The Life of Fishes.* Cleveland: World Publishing, 1966.

Smith, H. H. *Brazil.* 1879. Tale credited to Mario dos Rios, of Santarém.

147

Chapter Three

Darling, Lois and Louis. *Bird*. Boston: Houghton Mifflin (Riverside Press, Cambridge), 1962.

Eimerl, Sarel, and Devore, Irven. *The Primates*. New York: Time-Life, 1965.

Howell, F. Clark. *Early Man*. New York: Time-Life, 1970.

Laycock, George. *Strange Monsters & Great Searches*. Garden City: Doubleday, 1973.

Napier, John. *Bigfoot*. New York: Dutton, 1973.

Peissel, Michel. "Mustang, Nepal's Lost Kingdom." *National Geographic* 128, no. 4 (October 1965).

Rawicz, Slavomir. *The Long Walk*. New York: Harper & Row, 1956.

Stonor, Charles. *The Sherpa and the Snowman*. London: Hollis & Carter, 1955.

Von Nebesky-Wojkowitz, René. *Where the Gods Are Mountains*. London: Weidenfield & Nicolson, 1956.

Chapter Four

Eiseley, Loren. *The Immense Journey*. New York: Random House, 1957.

Green, John. *On the Track of Sasquatch*. Agassiz, B.C.: Cheam, 1971.

———. *The Year of the Sasquatch*. Agassiz, B.C.: Cheam, 1970.

Roosevelt, Theodore. *Wilderness Hunter*. New York, 1892.

Simons, E. L., and Ettel, P. E. "Gigantopithecus."

Stevens, James. *Paul Bunyan*. New York: Knopf, 1925.

148

D'Aulaire, Ingri and Edgar. *Norse Gods and Giants.* Garden City: Doubleday, 1967.

De Regniers, Beatrice Schenk. *The Giant Book.* New York: Atheneum, 1969.

Goodrich, Norma Lorre. *Ancient Myths.* New York: New American Library, 1960.

Hooijer, D. A. *Some Notes on the* Gigantopithecus *Question.* Leiden, Netherlands: Rijkmuseum of Natural History, 1949.

Lim Sian-tek. *Folk Tales from China.* New York: John Day, 1944.

Nahm, Milton C. *Selections from Early Greek Philosophy.* New York: Appleton-Century-Crofts, 1962.

Ross, Nancy Wilson. *Three Ways of Asian Wisdom.* New York: Simon and Schuster, 1966.

Teilhard de Chardin, Pierre. *The Appearance of Man.* New York: Harper & Row, 1965.

————, and Pei Wen-chung. "The Fossil Mammals from Locality of Choukoutien." *Pal. Sinica,* n.s., ser. C., no. 11 (1941).

Index

150

151

153

P2